South-Western
QUICK *START*

GRAPHING CALCULATOR APPLICATIONS
for Pre-Algebra

Claudia R. Carter

Mississippi School for Mathematics and Science
Columbus, Mississippi

Susanne K. Westegaard

Montgomery-Lonsdale Public School
Montgomery, Minnesota

South-Western Publishing Co.

ISBN: 0-538-62491-4

1 2 3 4 5 6 7 8 9 10 MA 02 01 00 99 98 97 96 95 94 93

Printed in the United States of America

Managing Editor: *Eve Lewis*
Developmental Editor: *Enid M. Nagel*
Senior Production Editor: *Mary Todd*
Coordinating Editor: Patricia *Matthews Boies*
Production Artist: *Sophia Renieris*
Marketing Manager: *Carol Ann Dana*
Mathematics Consultant: *William E. Royalty*

Table of Contents

Correlations

Correlation of *Quick Start Graphing Calculator Applications for Pre-Algebra* with *South-Western Pre-Algebra, Math Matters Book 1* and *Math Matters Book 2*

Quick Start Graphing Calculator Applications for Pre-Algebra	South-Western Pre-Algebra	Math Matters Book 1	Math Matters Book 2
Lessons	**Lessons**	**Lessons**	**Lessons**
1 Getting Started	——	——	——
2 Order of Operations	1-7	2-8	1-1
3 Adding Integers	2-2	6-2	2-2
4 Subtracting Integers	2-3	6-3	2-2
5 Multiplying Integers	2-4	6-4	2-3
6 Dividing Integers	2-5	6-5	2-3
7 Exponents and Integers	2-6	6-5	2-8
8 Fraction and Decimal Patterns	3-1,3-2	4-3	2-5
9 Fraction and Decimal Operations	3-3,3-4,3-5 3-6,3-7,3-8	4-5,4-6	2-5
10 Evaluating Variable Expressions	5-1	2-2	1-1
11 Investigating Percents	8-9	9-8	1-6
12 Perimeter and Area Investigation	9-4,9-5,9-6	5-4,11-2,11-4	6-2,6-4
13 Square Roots and the Pythagorean Theorem	10-4	11-6	6-1
14 Area and Circumference of a Circle	10-7	5-6,11-8	6-5
15 Graphing Equations	12-1	14-3	8-2,8-3,8-4
16 Solving Equations by Graphing	12-1	7-5	5-3
17 The Slope of a Line	12-2,12-3,12-4	14-4	9-1
18 Graphing Systems of Linear Equations	12-5	——	9-6
19 Graphing Inequalities: A Two-Dimensional View	6-9,6-10	7-11	5-6,8-8
20 Finding Averages and Drawing Scatter Plots	13-7	1-9	4-4,4-7
21 Exploring Data and Line Relationships	13-7	——	4-7

Name _____

Getting Started

Date _____

Objective You will become familiar with how to operate a graphing calculator.

Keys

■ Graphing calculators can draw graphs as well as perform mathematical calculations. Specific directions for using the *TI-81* and the *fx-7700G* are given in each lesson. These directions should provide you with enough information to make adaptations to other models of graphing calculators.

Look carefully at your calculator and notice the positions of the keys. Each key has a label on the key and labels above the key. When you press a key, the function shown by the label on the key will occur. In this book, the labels on the key are shown by ⎣ + ⎤.

When referring to the label above the key, the label is printed in brackets, for example [OFF]. These functions are activated by pressing the key ⎣SHIFT⎤ or ⎣2ND⎤ and then the key you want. Other labels above the keys are activated by pressing ⎣ALPHA⎤ and the key you want. Notice that the labels above the keys are colored the same as the corresponding shift key.

fx-7700G	TI-81
⎣AC⎤ is used to turn the calculator on and to clear the screen of the calculator. ⎣SHIFT⎤ [OFF] is used to turn the calculator off. ⎣EXE⎤ is used to execute a calculation.	⎣ON⎤ is used to turn the calculator on. ⎣CLEAR⎤ is used to clear the screen of the calculator. ⎣2ND⎤ [OFF] is used to turn the calculator off. ⎣ENTER⎤ is used to enter a calculation.

■ Most graphing calculators have menus. When you press a key that accesses a menu, several choices will appear on the screen of the calculator.

fx-7700G	TI-81
The function keys ⎣F1⎤ – ⎣F6⎤ are used to make menu selections. The label above each function key is active unless menu selections appear along the bottom of the screen. These selections appear when special keys such as ⎣SHIFT⎤ [DISP] are pressed.	Menu selections appear when special keys such as ⎣MODE⎤ are pressed. To make a menu selection, use the arrow keys ⎣◄⎤, ⎣►⎤, ⎣▲⎤, and ⎣▼⎤, to highlight your selection, then press ⎣ENTER⎤.

■ The mode menu is used to set special features. You should check the mode menu before you begin.

fx-7700G

Press `MDISP`.The following should be displayed.

```
      RUN/ COMP
  G-type: REC/CON
   angle: Deg
 display: Nrm1
```

To obtain the correct settings, use the Mode key.

`MODE` `1` for run (RUN).

`MODE` `+` for computation (COMP).

`MODE` `SHIFT` `+` for rectangular coordinates (REC).

`MODE` `SHIFT` `5` for connect (CON).

`SHIFT` [DISP] `F3` `EXE` for norm 1 (Nrm1).

TI-81

Press `MODE`.The following should be highlighted.

```
NORM Sci Eng
FLOAT 0123456789
Rad DEG
FUNCTION Param
CONNECTED Dot
SEQUENCE Simul
GRID OFF Grid on
RECT Polar
```

To make a correction, use the arrow keys to highlight your selection, then press `ENTER`. To leave mode, press `2ND` [QUIT]

■ It is not necessary to retype when you have made a typing error or want to change an entry. Both the *fx-7700G* and the *TI-81* have insert and delete keys.

fx-7700G

`DEL` deletes character at cursor location. `SHIFT` [INS] allows you to insert characters at the cursor location. `◀` `▶` `▲` `▼` are used to position cursor. `◀` is used to replay the previous line so you can edit it.

TI-81

`DEL` deletes character at cursor location. `INS` allows you to insert characters at the cursor location. `◀` `▶` `▲` `▼` are used to position cursor. `▲` is used to replay the previous line so you can edit it.

If you need more information about how to operate your calculator, be sure to consult your calculator's manual. Now you are ready to use your calculator to help you learn mathematics.

Name _____

Order of Operations

Date _____

Objective You will be able to evaluate mathematical expressions.

Keys fx-7700G [+] [−] [X] [÷] TI-81 [+] [−] [X] [÷]

■ When evaluating expressions involving more than one operation, you must use the rules for the order of operations.

1. Perform the operations in parentheses first.
2. Multiply or divide in the order they appear.
3. Add or subtract in the order they appear.

The graphing calculator also uses these rules for the order of operations with one exception. Implied multiplication will be calculated before any multiplication, division, addition, or subtraction that is shown by an operation symbol.

fx-7700G	TI-81

Example A 12 ÷ 4 x 3

12 [÷] 4 [X] 3 [EXE]

Answer: 9

12 ÷ 4 x 3

12 [÷] 4 [X] 3 [ENTER]

Answer: 9

Example B 12 ÷ 4 (3)

12 [÷] 4 [( ] 3 [ )] [EXE]

Answer: 1

12 ÷ 4 (3)

12 [÷] 4 [( ] 3 [ )] [ENTER]

Answer: 1

Notice that in Example A, the division is performed first and then the multiplication. However, in Example B the multiplication is performed first and then the division. When using a graphing calculator be sure to use the operation symbol when necessary so you will get the correct answer.

Exercises

Describe the correct order of operations for each. Then evaluate.

1. 42 ÷ 14 x 12 ÷ 9 = _____

_____

_____

2. 3 (8 − 4) + 6(15 − 8) ÷ 3 = _____

_____

_____

3. 3 ÷ 8 + 6(15 − 8) = _____

_____

_____

4. (29 − 5) ÷ (18 ÷ (9 − 3) x 2) = _____

_____

_____

Evaluate.

5. 48 ÷ 12 + 4 _____

6. 48 ÷ (12 + 4) _____

7. 48 + 12 ÷ 4 _____

8. (48 + 12) ÷ 4 _____

9. 9 − 2 x 3 x 2 − 3 _____

10. 9 − 2 x (3 x 2 − 3) _____

11. (9 − 2) x 3 x (2 − 3) _____

12. (((9 − 2) x 3) x 2) − 3 _____

13. 8 x 24 + 6 − 9 ÷ 3 _____

14. 8 x (24 + 6) − (9 ÷ 3) _____

15. 8 x (24 + 6 − 9) ÷ 3 _____

16. (8 x 24) + (6 − 9) ÷ 3 _____

Insert parentheses if needed to obtain the given result.

17. 16 ÷ 4 + 2 x 8 = 20

18. 16 ÷ 4 + 2 x 8 = $21\frac{1}{3}$

19. 16 ÷ 4 + 2 x 8 = $\frac{1}{3}$

20. 16 ÷ 4 + 2 x 8 = 0.8

21. 16 ÷ 4 + 2 x 8 = 48

Drawing Conclusions

Look at the answers to Exercises 1 – 8.

22. When an expression includes addition and subtraction, which operation do you perform first? _____

23. When an expression includes division and multiplication, which operation do you perform first? _____

24. When an expression includes addition and multiplication, which operation do you perform first? _____

25. When an expression includes division and addition, which operation do you perform first? _____

26. In what order should the following operations in an expression be performed?

multiplication _____ addition in parentheses _____ subtraction _____

Challenge

27. *Columbus sailed the ocean blue in fourteen hundred ninety-two, or was it Columbus sailed the deep blue sea in fourteen hundred ninety-three?*

Using the digits in 1492, the operations +, −, ÷, x, and parentheses, find expressions that give the values of 1 through 10.

For example, using the digits of 1493:

9 − (3 + 4 + 1) = 1   9 − (4 + 3) ÷ 1 = 2   9 − 4 − 3 + 1 = 3

Name _____

Adding Integers

Date _____

Objective   You will be able to add integers.

Keys   **fx-7700G**    **TI-81**

■ When adding integers you will need to enter the sign of negative integers. Note that the negative key [(−)] and the minus key [−] are different keys on the *TI-81*. The *fx-7700G* uses [−] for subtraction as well as negation. When entering an exercise that starts with a negative number, you will need to clear the screen using [AC] on the *fx-7700G*.

fx-7700G	TI-81

Example A   Find the sum of 2 and −5.   Find the sum of 2 and −5.

Answer: −3   *Answer:* −3

Example B   Find the sum of −7 and −3.   Find the sum of −7 and −3.

[AC][−] 7 [+][−] 3 [EXE]   [(−)] 7 [+][(−)] 3 [ENTER]

Answer: −10   *Answer:* −10

CAUTION: Clear the screen of the fx-7700G between exercises.

Exercises

Find each sum.

1. 12 + 13   _____   **2.** (−12) + 13   _____

3. (−6) + (−3)   _____   **4.** (−9) + (−17)   _____

5. (−13) + 19   _____   **6.** 7 + 4   _____

7. (−24) + 25   _____   **8.** (−14) + 32   _____

9. (−7) + 13   _____   **10.** (−8) + (−3)   _____

11. 36 + (−29)   _____   **12.** (−96) + 100   _____

13. (−369) + 369   _____   **14.** (−16) + (−16)   _____

15. (−200) + 201   _____   **16.** 6000 + (−6001)   _____

17. (−13) + 24 + (−11)   _____   **18.** (−12) + (36) + (−9)   _____

LESSON 3 Adding Integers

Drawing Conclusions

Use your answers from Exercises 1-18 to complete each of the following sentences.

19. When adding only positive integers, the sign of the answer is _____.

20. When adding only negative integers, the sign of the answer is _____.

21. When adding positive and negative integers, the sign is determined by _____

_____

22. When adding opposite integers (like –8 and 8), the sum is _____.

Applications

A matrix is a rectangular array of numbers. To add matrices (plural of matrix), add the numbers (elements) in corresponding positions.

Example:
$$\begin{bmatrix} -3 & -5 \\ 4 & 3 \end{bmatrix} + \begin{bmatrix} 3 & 6 \\ -3 & 7 \end{bmatrix} = \begin{bmatrix} -3+3 & -5+6 \\ 4+(-3) & 3+7 \end{bmatrix} = \begin{bmatrix} 0 & 1 \\ 1 & 10 \end{bmatrix}$$

Add the following matrices.

23.
$$\begin{bmatrix} -6 & 4 & -3 \\ 16 & -11 & -19 \\ 8 & -9 & -17 \end{bmatrix} + \begin{bmatrix} 8 & -9 & 3 \\ -9 & 12 & -20 \\ 3 & 8 & 13 \end{bmatrix} = \begin{bmatrix} \_\_ & \_\_ & \_\_ \\ \_\_ & \_\_ & \_\_ \\ \_\_ & \_\_ & \_\_ \end{bmatrix}$$

24.
$$\begin{bmatrix} 7 & -1 & 10 \\ -5 & 14 & 2 \\ 11 & -8 & -15 \end{bmatrix} + \begin{bmatrix} -3 & -4 & 5 \\ 11 & -14 & -7 \\ -4 & 7 & -5 \end{bmatrix} = \begin{bmatrix} \_\_ & \_\_ & \_\_ \\ \_\_ & \_\_ & \_\_ \\ \_\_ & \_\_ & \_\_ \end{bmatrix}$$

Solve each problem.

25. Your football team starts on the 30 yard line of the opposing team. On the first down, you gain 8 yards. On the second down, you lose 6 yards. On the third down, you gain 7 yards. If you have not gained at least 10 yards, your team must kick the ball away. Do you kick the ball? If so, from what yard marker?

26. The thermometer in the classroom started at 68 degrees. After the first hour, it dropped 4 degrees; after the second hour, it dropped 2 degrees. During the third hour, the temperature remained the same. Then it rose 5 degrees during the next hour. What was the temperature after the fourth hour?

27. Write your own problem that can be solved by adding integers. Use at least three integers.

Name _____

Date _____

Subtracting Integers

Objective You will be able to subtract integers.

Keys fx-7700G    TI-81

■ For subtraction you use the  key on both the *fx-7700G* and the *TI-81*. When subtracting integers you will need to enter the sign of negative integers. Remember that the negative key is  on the *TI-81*.

fx-7700G	TI-81

Example A Subtract 5 from –3.

[AC] [–] 3 [–] 5 [EXE]

Answer: –8

Subtract 5 from –3.

[(–)] 3 [–] 5 [ENTER]

Answer: –8

Example B Subtract –7 from –4.

[AC] [–] 4 [–] [–] 7 [EXE]

Answer: 3

Subtract –7 from –4.

[(–)] 4 [–] [(–)] 7 [ENTER]

Answer: 3

CAUTION: Clear the screen of the fx-7700G between exercises.

Exercises

Subtract.

1. 12 – 13 _____

2. (–12) – 13 _____

3. (–6) – (–3) _____

4. (–9) – (–17) _____

5. (–13) – 19 _____

6. 7 – 4 _____

7. 24 – 36 _____

8. (–14) – 32 _____

9. (–7) – 13 _____

10. (–8) – (–3) _____

11. 35 – (–29) _____

12. (–96) – 100 _____

13. 369 – 360 _____

14. (–16) – (–16) _____

15. (–200) – 201 _____

16. 6000 – (–6001) _____

17. (–12) – 13 – (–24) _____

18. 24 – (–13) – 14 _____

Drawing Conclusions

19. When adding opposite integers (like –23 and 23), the sum is _____.

20. Look at the patterns in these problems and make some conjectures.

(–6) + 4 = _____	and	(–6) – (–4) = _____
7 + (–3) = _____	and	7 – 3 = _____
6 + 3 = _____	and	6 – (–3) = _____
(–3) + (–2) = _____	and	(–3) – 2 = _____

Applications

21. Your bank balance is $123.41. You write checks for $25.76, $36.75, and $65.23. Do you have a balance or are you overdrawn? What is the amount of balance or overdraft?

22. The high temperature in Amarillo, Texas was 108°F. The low temperature in Minneapolis, Minnesota was –60°F. What was the difference between these temperatures?

23. RESEARCH ▼
For one week, follow one stock in the newspaper. Record the daily changes.
At the end of the week, determine if the stock had a weekly gain or loss.

Monday _____   Tuesday _____   Wednesday _____

Thursday _____   Friday _____   Gain or Loss _____

24. In a magic square the sum of every row, every column, and both diagonals is the same. Is this a magic square?

16	1	–2
–13	5	23
12	9	–6

____

25. Add –7 to each entry in the square in Exercise 24. Find the sum of each row, column, and diagonal. Is it a magic square?

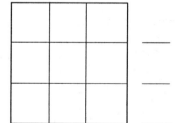

____ ____ ____

____ ____ ____

Name _____

Date _____

Multiplying Integers

Objective   You will be able to multiply integers.

Keys   fx-7700G $\boxed{\text{X}}$ TI-81 $\boxed{\text{X}}$

■ For multiplication you use the $\boxed{\text{X}}$ key on both the *fx-7700G* and the *TI-81*. Remember to use the correct key for negative integers.

fx-7700G	TI-81
Example A Multiply –2 by 5.	Multiply –2 by 5.

Example A

fx-7700G: $\boxed{\text{AC}}$ $\boxed{-}$ 2 $\boxed{\text{X}}$ 5 $\boxed{\text{EXE}}$

Answer: –10

TI-81: $\boxed{(-)}$ 2 $\boxed{\text{X}}$ 5 $\boxed{\text{ENTER}}$

Answer: –10

Example B   Multiply –3 by –13.

fx-7700G: $\boxed{\text{AC}}$ $\boxed{-}$ 3 $\boxed{\text{X}}$ $\boxed{-}$ 13 $\boxed{\text{EXE}}$

Answer: 39

TI-81: $\boxed{(-)}$ 3 $\boxed{\text{X}}$ $\boxed{(-)}$ 13 $\boxed{\text{ENTER}}$

Answer: 39

CAUTION:
With the TI-81 remember to use [(–)] for negative integers.

Exercises

Subtract.

1. –6 x (–5)   _____

2. –7 x 2   _____

3. –14 x 17   _____

4. 21 x 23   _____

5. 23 x (–16)   _____

6. –432 x (–732)   _____

7. 17 x (–203)   _____

8. 1232 x (–4)   _____

9. –435 x 345   _____

10. –234 x 234   _____

11. –173 x (–173)   _____

12. 10002 x (–5)   _____

13. –17 x 13 x (–4)   _____

14. –5 x (–134) x (–2)   _____

15. –5 x (–5) x (–5)   _____

16. –1 x (–1) x (–1) x (–1)   _____

Drawing Conclusions

Use your answers from Exercises 1-16 to complete each of the following.

17. When multiplying two positive integers, the sign of the answer is

_____ .

18. When mutliplying a negative integer by a positive integer, the sign of the

answer is _____ .

19. When multiplying two negative integers, the sign of the answer is _____ .

20. Examine Exercises 15 and 16, draw a conclusion about the signs of the answers.

21. Jim wants you to add these numbers: –6, –6, –6, –6, –6, –6, and –6. Can you explain a way to save keystrokes on the calculator?

Applications

22. A scalar for a matrix is a number in front of the matrix. Each entry of the matrix must be multiplied by the scalar. Multiply a scalar of –3 times this matrix.

$$-3 \quad \times \quad \begin{bmatrix} -12 & 13 & -5 \\ 14 & -9 & 0 \\ -16 & 24 & -9 \end{bmatrix} = \begin{bmatrix} \underline{} & \underline{} & \underline{} \\ \underline{} & \underline{} & \underline{} \\ \underline{} & \underline{} & \underline{} \end{bmatrix}$$

23. In Montgomery, Minnesota the temperature was 24 degrees below zero. Frank, from Fargo, North Dakota calls Mary in Montgomery and says the thermometer reading was twice as much in Fargo. What was the reading in Fargo?

24. The price of Burger Delight stock was $39 per share. The price fell $1.25 each day over a 5-day period. What was the total change in price? What was the price of Burger Delight stock after the drop?

Name _____

Dividing Integers

Date _____

Objective   You will be able to divide integers.

Keys   fx-7700G [÷]   TI-81 [÷]

■ For division you use the [÷] key on both the *fx-7700G* and the *TI-81*.

fx-7700G	TI-81
Example A Divide –3520 by 4.	Divide –3520 by 4.
[AC] [–] 3520 [÷] 4 [EXE]	[(–)] 3520 [÷] 4 [ENTER]
Answer: –880	*Answer:* –880
Example B Divide –5280 by –12.	Divide –5280 by –12.
[AC] [–] 5280 [÷] [–] 12 [EXE]	[(–)] 5280 [÷] [(–)] 12 [ENTER]
Answer: 440	*Answer:* 440

Exercises

Find each quotient.

1. –928 ÷ (–16)   _____

2. –4125 ÷ (–125)   _____

3. 5562 ÷ (–18)   _____

4. –1344 ÷ 84   _____

5. –1000 ÷ 0.08   _____

6. –1836 ÷ (–36)   _____

7. –500 ÷ 25   _____

8. –7280 ÷ (–80)   _____

9. –2160 ÷ (–3) ÷ 5 ÷ (–8) ÷ (–3) _____

10. –3240 ÷ (–2) ÷ (–3) ÷ (–5) ÷ (–6) _____

Drawing Conclusions

Use the answers to 1-10 to complete each sentence.

11. The sign of the quotient of two positive integers is _____ .

12. The sign of the quotient of two negative integers is _____ .

13. The sign of the quotient of a negative and a positive integer is _____ .

14. Look at Exercises 9 and 10. A list of divided integers has a negative

result if _____

and a positive result if _____ .

LESSON 6 Dividing Integers

Investigation

Find the answers to the following exercises:

15. $-6720 \div (-8) \times 5 \div (-6) \div (-4) =$ _____

16. $4480 \div (-16) \times 2 \div (-5) \div (-8) =$ _____

17. $-1890 \div (-9) \div (-5) \times (-6) \div (-7) =$ _____

18. Look back at your answer to Exercise 14. Do your answers for 15 - 17 follow your rule?

Applications

19. A class of 25 students is going to take a field trip to the Science Museum. The cost of transportation, food, and admission tickets is $300. If each student pays an equal share, how much does each student owe?

20. In Exercise 19, suppose three students have to stay home; then how much would each student owe? Explain your reasoning.

21. In Exercise 19, suppose the class has a fund raiser to earn money for the class trip which raises $125. How will this change the answers to 19 and 20?

22. A deep sea diving capsule dove to a depth of 5290 feet. As it ascends, it needs to stop every 1000 feet for a half hour to adjust the pressure. How many stops will be needed before reaching sea level? Explain your answer.

23. Find the average of these integers: $-92, -75, 32, -17$, and 47.

24. Find five integers, with at least one positive, that have an average of -15. Explain your reasoning.

Name _____

Date _____

Exponents and Integers

Objective   You will be able to evaluate expressions containing exponents and integers.

Keys   fx-7700G $\boxed{x^y}$ TI-81 $\boxed{\wedge}$

■ Numbers written in *exponential form* include a base and an exponent. For example, 3^5 is written in exponential form where 3 is the base and 5 is the exponent. The exponent tells you how many times the base is used as a factor. To enter exponents on a graphing calculator, you need to use the exponent key. For the *fx-7700G*, you use $\boxed{x^y}$ and for the *TI-81*, you use $\boxed{\wedge}$.

fx-7700G	TI-81

Example A   Evaluate $(-5)^6$.   Evaluate $(-5)^6$.

Answer: 15,625   *Answer:* 15,625

Example B   Evaluate $5^2 - (-3)^4$.   Evaluate $5^2 - (-3)^4$.

Answer: –56   *Answer:* –56

Exercises

Evaluate.

1. $(-4)^7$ _____   2. 4^9 _____

3. 5^{11} _____   4. $(-1)^{234}$ _____

5. 12^4 _____   6. $(-6)^{10}$ _____

7. $(-9)^8$ _____   8. 7^8 _____

9. $(-8)^7$ _____   10. $(-8)^8$ _____

11. $(-2)^9$ _____   12. $(-2)^{10}$ _____

13. $(-3)^6$ _____   14. $(-3)^7$ _____

15. $(-3)^3 - 5^2$ _____   16. $(-7 + 5)^4 \div (-2)^3$ _____

17. $12^2 - (4 - 4 \times 2)^2$ _____   18. $12^2 + (-3)^3 - 5^2$ _____

19. $2(18 \div (-3) - (-2))^3$ _____   20. $-11 + 4^3 + (-3)^2 - 2^4$ _____

21. $4^4 \div (-2)^3 \div (-1)^3$ _____   22. $(5^2 - 8) + 9 \times 7$ _____

Drawing Conclusions

Use your answers from Exercises 1–14 to complete each of the following sentences.

23. When a negative integer is raised to an even power, the sign of the answer

is _____.

24. When a negative integer is raised to an odd power, the sign of the answer

is _____.

25. Evaluate each of the following.

$-(5)^3$ _____ $(-5)^3$ _____ -5^3 _____

$-(5)^6$ _____ $(-5)^6$ _____ -5^6 _____

Write about the results. Can you see any reason to use parentheses?

Applications

A formula for interest is

$I = Prt$

where I is the interest earned, P is the amount invested, r is the rate as a decimal, and t is the time in years. Remember that 8% can be written as the decimal 0.08.

26. Determine the interest earned if $1000 is invested at a rate of 8% for 10 years.

27. Determine the interest earned if $10,000 is invested at a rate of 8% for 10 years.

28. Determine the interest earned if $100,000 is invested at a rate of 8% for 10 years.

29. Is there a pattern? Predict what would happen if $1,000,000 were invested at the same rate for the same time period.

30. Complete the cross number puzzle on the following page.

Across

1. $-600 \div 12$
2. $14 \times (-5)$
3. $-7 + (-300) + 42$
6. $-151{,}755 \div (-15)$
8. $-1750 - 14$
9. $-19 - (-32)$
10. $2 \times (-16 + 5)$
11. $-9 - (-11) + 4 - 6$
12. $-21{,}600 \div (-6) \div (-5) \div (-9)$
13. $8^2 - (-10)^2$
14. $(-2)^8$
16. $-7 - (-14)$
17. $-2 \times (-5) \times (-9) \times (-7)$
19. $-850 \div 17$
20. $6 \times (-13 + (-17))$
21. $(-9)^2$
22. -3×18
23. $(-7 + (-3)) \times (-5 + 3) \div 2$
24. $-6 + 7 \times -3 + 370$
26. Start at sea level, descend 40 ft, ascend 10 ft, descend 35 ft
27. Start with $400, spend $250, deposit $161, spend $92
28. $-13 - (-7) - 5$
29. $-360 \div (-2) \div (-3) \div (-4) \div (-5)$
30. $-12 \times (-5 + 6)$
31. $(-2 + 12) + (7)^3$
33. $-36 \div (-8) \times 4$
34. $-16 - (-16)$

Down

1. -17×3
2. $-600 + (-110) + (-3)$
3. $(-3)^3$
4. $-66 \times (-100) + 5$
5. $-2 \times (-3)^3$
7. $-(-11)^1$
8. $(-2)^2 \times (-3)$
10. $-13 - 13$
12. $((-3)^3 \times (29)) \div 9$
13. $-2 \times (-3) \times (-5)$
14. $(-2)^5 + 52$
15. $-15 \times (-40) + (-7) \times (-2)$
17. $(-17) \times (-2)^2 \times (-1)$
18. $-9000 \div (-3) - (-69) - 38$
19. $(-2)^4 \times (-3) - 3$
20. $(-10)^6 \div (-10)^5$
21. $(-2)^4 \times 5 - (-3)$
22. $5 \times (-11)$
23. $(-2)^4$
24. $(2)^5 - 1$
25. $(5)^3 \times (-2)^2 - 7$
26. $-(2)^6 + 3$
27. $(-15)^2$
28. $(-13)^7 \div (-13)^6$
29. $(-3)^5 + (-5)^3 + (-2)^3 - (-2)^2$
30. $2 \times (-2)^3$
32. $-16{,}575 \div 25 \div (-13)$

Name _____

Date _____

Fraction and Decimal Patterns

Objective   You will be able to write fractions as decimals.

Keys   fx-7700G $\boxed{\div}$ TI-81 $\boxed{\div}$

■ On a graphing calculator you can divide to write a fraction as a decimal. Notice that the calculator gives a repeating decimal rounded to 10 digits, since the calculator can only display 10 digits. However, you should be able to see the repeating pattern. Remember that you can write a repeating decimal like 0.333333... as $0.\overline{3}$.

fx-7700G	TI-81

Example A   Write $\frac{1}{3}$ as a decimal.

$1 \boxed{\div} 3 \boxed{\text{EXE}}$

Answer: 0.3333333333 or $0.\overline{3}$

Write $\frac{1}{3}$ as a decimal.

$1 \boxed{\div} 3 \boxed{\text{ENTER}}$

Answer: 0.3333333333 or $0.\overline{3}$

Example B   Write $\frac{1}{6}$ as a decimal.

$1 \boxed{\div} 6 \boxed{\text{EXE}}$

Answer: 0.6666666667 or $0.\overline{6}$

Write $\frac{1}{6}$ as a decimal.

$1 \boxed{\div} 6 \boxed{\text{ENTER}}$

Answer: 0.6666666667 or $0.\overline{6}$

Exercises

Write a decimal for each fraction. Remember to notice repeating decimals.

1. $\frac{1}{2}$ = _____

2. $\frac{3}{8}$ = _____

3. $\frac{2}{5}$ = _____

4. $\frac{2}{3}$ = _____

5. $\frac{5}{4}$ = _____

6. $\frac{1}{8}$ = _____

7. $\frac{5}{6}$ = _____

8. $\frac{5}{11}$ = _____

9. $\frac{11}{7}$ = _____

10. $\frac{1}{5}$ = _____

11. $\frac{3}{7}$ = _____

12. $\frac{10}{27}$ = _____

Drawing Conclusions

Let's look more closely at specific fractions and their decimal patterns. Write a decimal for each of these fractions.

13. $\frac{1}{9}$ = _____

14. $\frac{2}{9}$ = _____

15. $\frac{3}{9}$ = _____

16. $\frac{13}{99}$ = _____

17. $\frac{37}{99}$ = _____

18. $\frac{67}{99}$ = _____

19. From the results of Exercises 13 - 18, write a rule about dividing by nine.

_____

_____

Write each decimal value. Can you tell what number is the remainder by comparing these answers to those in Exercises 1 - 9? Write the final answer as a mixed number.

	Decimal Value	Remainder	Mixed Number
20. $\frac{10}{7}$	_____	_____	_____
21. $\frac{100}{9}$	_____	_____	_____
22. $\frac{241}{3}$	_____	_____	_____
23. $\frac{256}{5}$	_____	_____	_____
24. $\frac{171}{8}$	_____	_____	_____

Write the decimal values for each.

25. $\frac{2}{7}$ = _____

26. $\frac{4}{7}$ = _____

27. $\frac{5}{7}$ = _____

28. $\frac{6}{7}$ = _____

29. Look closely at the digits that repeat in Exercises 25 - 28. What pattern do you see?

30. RESEARCH ▼

Investigate with your calculator to see how many other number patterns you can find.

Name _____

Date _____

Fraction and Decimal Operations

Objective   You will be able to perform operations with fractions and decimals.

Keys   **fx-7700G** [a b/c]

■ To perform operations with mixed numbers you need to enter the mixed number in the calculator correctly. Remember that $5\frac{5}{6}$ means $5 + \frac{5}{6}$.

fx-7700G	TI-81

Example A   Write $5\frac{5}{6}$ as a decimal.

5 [+] 5 [÷] 6 [EXE]

Answer: 5.833333333 or 5.8$\overline{3}$

Write $5\frac{5}{6}$ as a decimal.

5 [+] 5 [÷] 6 [ENTER]

Answer: 5.8333333333 or 5.8$\overline{3}$

When performing operations with fractions or mixed numbers be sure to use parenthesis when necessary. When multiplying with a mixed number, use parentheses around the mixed number so the addition will be performed first.

fx-7700G	TI-81

Example B   Multiply $4\frac{2}{3}$ by 21.

21 [EXE]

Answer: 98

Multiply $4\frac{2}{3}$ by 21.

21 [ENTER]

Answer: 98

Some calculators such as the *fx-7700G* have a special key for entering fractions [a b/c]. To enter a fraction such as $\frac{4}{5}$, press 4 [a b/c] 5 [EXE]. You will see 4⌐5 on your screen. The ⌐ represents the fraction bar. Press [a b/c] to get the decimal value of the fraction, 0.8. If you enter an improper fraction such as $\frac{20}{3}$, the calculator will automatically reduce the fraction giving you 6⌐2⌐3, that is $6\frac{2}{3}$.

To enter a mixed number such as $4\frac{2}{3}$, press 4 [a b/c] 2 [a b/c] 3 [EXE]. You will see 4⌐2⌐3.

When you enter fractions with the fraction key, you do not need to use parentheses when multiplying with a mixed number. For Example B, enter

4 [a b/c] 2 [a b/c] 3 [x] 21 [EXE].

Exercises

Complete the cross-number puzzle. When entering decimal numbers do not include the decimal point or any leading zero. When entering negative numbers do not include the negative sign.

ACROSS

1. $5\frac{5}{6} \times 36 = ?$

3. $3\frac{1}{7} \div \frac{1}{49} = ?$

6. $7\frac{5}{12} + 8\frac{1}{2} + 6\frac{1}{4} + 11\frac{5}{6} = ?$

7. Decimal equivalent for $\frac{21}{40} = ?$

9. The LCD for $\frac{1}{3} - \frac{6}{7} + \frac{8}{11}$ is ?

11. $5\frac{3}{4} \div \left(-\frac{1}{4}\right)$

12. The reciprocal of $\left(-\frac{2}{3} - \left(-\frac{1}{2}\right)\right) \div \left(-\frac{3}{2} - \frac{3}{2}\right)$

13. $\left(\frac{23}{5}\right) \times 5$

14. $-17\frac{1}{4} \times \left(-23\frac{1}{8}\right) \div \frac{15}{32} + 62$

16. $5\frac{1}{4} - 6\frac{1}{3} + 8\frac{1}{9} - 6\frac{1}{36}$

DOWN

1. $\frac{2}{3} \times ? = 16$

2. 0.0534 rounded to the nearest thousandth is ?

3. $11 \div 5 = 2$ with remainder ?

4. $67\frac{5}{8} - 11\frac{1}{4} - 4\frac{3}{8} = ?$

5. Take the remainder of $\frac{144}{7}$, add it to the whole number part, then multiply by 18.

6. $\frac{21}{19} = \frac{?}{361}$

8. Round 0.5134 to the nearest hundredth.

10. $\left(-2\frac{1}{5}\right)\left(-\frac{2}{3}\right)\left(-7\frac{1}{5}\right)(-12) = ?$

17. 0.3468 – 0.2173 + 0.4317 – 0.5312

18. Write the fraction answer to clue 19 down. Put the numerator in the first box and the denominator in the second box.

19. The answer to (–5.3) x (–122.7) rounded to the nearest whole number

21. Take the denominator for $\left(\frac{1}{2}\right)^9$, then subtract 100.

24. Pablo can type 4.5 lines into the spreadsheet per minute. How many lines will he have typed after $16\frac{8}{9}$ minutes?

25. 121.7 + 168.5 – 502.3 + 213.1 = ?

27. If Clyda reads $\frac{5}{7}$ of a page per minute and she reads a book in 11.84 hours, approximately how many pages are in the book?

29. $\frac{3}{17} = \frac{6}{?}$

30. The reduced denominator after multiplying $\left(\frac{1}{4}\right)\left(\frac{3}{8}\right)\left(\frac{5}{6}\right)\left(\frac{18}{5}\right)$

31. $\frac{3}{2}$ of 12 dozen less two dozen

32. (–0.01)(0.7) answer shares something in common with a movie spy character.

11. If the answer to $-\frac{23}{14} + \frac{8}{7} - \frac{114}{5}$ is written as an improper fraction, the numerator is ?

15. 10.36 – 15.28 + 105.6 – (–4.82) = ?

16. The decimal digits for (0.3)(0.7)(0.8)(0.9) + 0.0022 are ?

19. A given nutty chocolate chip cookie recipe calls for $1\frac{1}{4}$ cup of pecans. If Dewayne wanted to halve the recipe, what decimal number would represent the cups of nuts?

20. The answer to 23.58 ÷ 302.6 rounded to the nearest ten thousandths

22. (0.51)³ rounded to four decimal places

23. The hundredths decimal number between 0.24 and 0.26

25. When dividing a number by 7, the two digits that always follow 7 in the answer are ?

28. What are the two digits that repeat for the decimal equivalent of $\frac{1}{99}$?

30. The reciprocal of $\frac{1}{6}$ times the reciprocal of $\frac{1}{5}$

Name _____

Date _____

Evaluating Variable Expressions

Objective You will be able to evaluate expressions for several values.

Keys fx-7700G [◄] [DEL] [SHIFT] [INS]   TI-81 [▲] [◄] [DEL] [INS]

■ You can use the editing features of the graphing calculator to evaluate variable expressions for several values of the variable. In Example A, you evaluate the expression by replacing the variable by the given value.

fx-7700G	TI-81

Example A Evaluate $-3y - 2$ if $y = -4$.

Answer: 10

Evaluate $-3y - 2$ if $y = -4$.

Answer: 10

In Examples B and C, the same expression as in Example A is evaluated for different values of the variable by using the editing keys. For the *fx-7700G*, use the left arrow [◄] to replay the previous expression and move to the point where you want to replace the original value. Then use the delete key [DEL] and the insert key [SHIFT] [INS] as needed. For the *TI-81*, use the up arrow [▲] to replay the previous expression and use the left arrow [◄] to move to the point where you want to replace the original value.

fx-7700G	TI-81

Example B Evaluate $-3y - 2$ if $y = 3$.

3 [EXE]

Answer: −11

Evaluate $-3y - 2$ if $y = 3$.

3 [ENTER]

Answer: −11

Example C Evaluate $-3y - 2$ if $y = 11$.

[◄] [◄] [◄] [◄] [DEL] [SHIFT]
[INS] 11 [EXE]

Answer: −35

Evaluate $-3y - 2$ if $y = 11$.

11 [ENTER]

Answer: −35

LESSON 10 Evaluating Variable Expressions

Exercises

Evaluate the given expression for the given values of the variable.

1. p $3p - (-5)$ **2.** q $-6q + 7$ **3.** w $-9(2w + 1)$

p		q		w	
-5	_____	-8	_____	-5	_____
-1	_____	-7.4	_____	-4.69	_____
0	_____	-3	_____	-1	_____
4	_____	2.5	_____	0	_____
11	_____	11	_____	3	_____
16.5	_____	17	_____	7	_____

Use the information from Exercises 1-3 to find values for p, q, and w that yield the following results.

4. What value for p yields 32? _____   **5.** What value for p yields -28? _____

6. What value for p yields 12.5? _____   **7.** What value for q yields 37? _____

8. What value for q yields 91.6? _____   **9.** What value for w yields 27? _____

10. What value for w yields -34.2? _____   **11.** What value for w yields -154.8? _____

Applications

The formula for changing Fahrenheit degrees to Celsius degrees is

$$C = \frac{5}{9}(F - 32)$$

Find each of the following to the nearest degree:

12. 75°F = _____ °C   **13.** 100°F = _____ °C

14. 50°F = _____ °C   **15.** 32°F = _____ °C

16. Use your answers to Exercises 12 - 15 to help you find 16°C = _____ °F.

The formula for the area of a triangle is $A = \frac{1}{2}bh$.

17. Find A if $b = 7$ and $h = 3$. _____   **18.** Find A if $b = 10$ and $h = 25$. _____

19. Find A if $b = 14$ and $h = 6$. _____   **20.** Find b if $h = 7$ and $A = 14$. _____

21. Find h if $b = 13$ and $A = 156$. _____

Evaluate.

22. $(a^2 + b^2)^2$ if $a = 2.7$ and $b = 1.74$ _____

23. $\frac{xy^2}{11}$ if $x = 74$ and $y = 16$ _____

24. $r + 3(s + 4t)$ if $r = 17.5$, $s = 7$, and $t = -13.1$. _____

Name_____

Date_____

Investigating Percents

Objective You will be able to solve problems involving percents.

Keys fx-7700G   TI-81 ◻ ◻ [ANS]

■ The formula for compound interest is

$$P\left(1 + \frac{r}{n}\right)^{nt}$$

where P is the amount invested, r is the rate of interest, n is how many times interest is calculated per year, and t is the number of years. Recall that percents should be written as decimals so that $7\frac{1}{2}\% = 0.075$. The editing capabilities of the graphing calculator are helpful when using the compound interest formula repeatedly. Use the replay key to retrieve the previous calculation, change any numbers desired, and recalculate as shown in the example. The answer feature uses the most recent answer. When n is 1 or t is 1 there is no need to enter these values.

fx-7700G	TI-81

Example

How long will it take $100 to double at $7\frac{1}{2}\%$ interest compounded annually?

$P(1 + r)$

100 [( ] 1 [+] 0.075 [ )] [EXE]

Answer: 107.5

Replay the previous calculation.
[◄]
Then press the left arrow [◄] until you are on the 1 of 100. Replace 100 with the answer to the calculation.

[ANS] [DEL] [DEL] [EXE]

Answer: 115.5625

Then repeat the calculation with the new answer.

[EXE]

Answer: 124.2296875

How long will it take $100 to double at $7\frac{1}{2}\%$ interest compounded annually?

$P(1 + r)$

100 [( ] 1 [+] 0.075 [ )] [ENTER]

Answer: 107.5

Replay the previous calculation.
[▲]
Then press the left arrow [◄] until you are on the 1 of 100. Replace 100 with the answer to the calculation.

[2ND] [ANS] [DEL] [DEL] [ENTER]

Answer: 115.5625

Then repeat the calculation with the new answer.

[ENTER]

Answer: 124.2296875

Continue pressing [EXE] or [ENTER]
until the answer is greater than $200.
Be sure to keep track of how many times
you press [EXE] or [ENTER]. After ten times
the amount is $206, so the answer is
10 years.

133.5469141
143.5629326
154.3301526
165.904914
178.3477826
191.7238662
206.1031562

Exercises

1. The numbers above the diagonal line represent numerators and those below represent denominators. Make a list of all the fractions that represent 25%, then shade in the corresponding squares. Each member of the pair can be used only once.

1	25	11	27	6	▓
5	3	24	34	▓	9
13	2	21	▓	18	196
32	15	▓	100	4	20
9	▓	30	84	81	8
▓	18	68	96	52	44

2. What symbol is suggested by the shading in Exercise 1? _____

3. Complete the magic square.

0.11	0.18	0	0.02	
	-0.01	0.01		0.15
-0.02		0.07		0.16
	0.06		0.20	
0.1	0.12	0.19		0.03

4. What percent is the sum for the magic square in Exercise 3? _____

Drawing Conclusions

5. Predict how long it will take for $100 to double at $7\frac{1}{2}$% interest if

    compounded quarterly. Verify with your calculator. Use $100\left(1 + \frac{0.075}{4}\right)^4$.

6. How long will it take $100 at $7\frac{1}{2}$% interest to triple compounded annually?
    Quarterly?

7. If the interest rate rises to 8%, how long does it take $100 to double and
    triple compounded annually? Quarterly?

Applications

8. Marvin wants to purchase a coat that originally listed for $117.95. If he can
    take an additional 20% off the price, already marked down 30%, how much
    will the coat cost Marvin before taxes?

9. In Exercise 8, would it be better for Marvin to have a 50% discount?

10. A chemical solution contains 40% acid in 56 ounces of solution. How many
    ounces are acid?

11. Within the next year, the city of Montegue wants to increase the number of
    townspeople who recycle from 12,000 to 14,000. If the population of
    Montegue is approximately 20,000 people, what percent of the population
    currently recycles?

12. In Exercise 11, what percent increase in participation is Montegue hoping
    for within the next year?

13. **RESEARCH ▼**
    Find out what percent of the population in your town or nearby city
    recycles. What items are recycled? What percent of total garbage does this
    represent?

Name _____

Date _____

Objective   You will be able to find the perimeter and area of a rectangle.

■ The *perimeter* is the distance around a closed figure. You can find the perimeter by adding the lengths of its sides.

Perimeter of a rectangle: $P = 2l + 2w$

The number of units needed to cover a figure is its *area*.

Area of a rectangle: $A = l \times w$

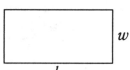

fx-7700G

Example   Find the perimeter and area of a rectangle with length 7 cm and width 5 cm.

$P = 2l + 2w = 2(7) + 2(5)$

2 ⌈ (⌉ 7 ⌈) ⌉ + 2 ⌈ (⌉ 5 ⌈) ⌉

Answer: 24 cm

$A = l \times w = 7 \times 5$

7 ⌈ x ⌉ 5

Answer: 35 cm²

TI-81

Find the perimeter and area of a rectangle with length 7 cm and width 5 cm.

$P = 2l + 2w = 2(7) + 2(5)$

2 ⌈ (⌉ 7 ⌈) ⌉ + 2 ⌈ (⌉ 5 ⌈) ⌉

Answer: 24 cm

$A = l \times w = 7 \times 5$

7 ⌈ x ⌉ 5

Answer: 35 cm²

Exercises

Find the perimeter and area of each rectangle.

1.

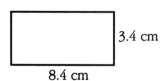

3.4 cm

8.4 cm

2.

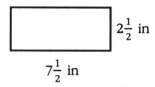

$2\frac{1}{2}$ in

$7\frac{1}{2}$ in

3.

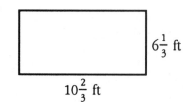

$6\frac{1}{3}$ ft

$10\frac{2}{3}$ ft

4.

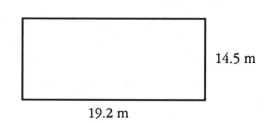

14.5 m

19.2 m

Drawing Conclusions

A rectangle has an area of 12 square units. Complete the chart to show possible widths. Round to the nearest hundredth.

	l	w	Perimeter		l	w	Perimeter
5.	1	12	2(1) + 2(12) = 26	**11.**	7	_____	_____
6.	2	6	_____	**12.**	8	_____	_____
7.	3	_____	_____	**13.**	9	_____	_____
8.	4	_____	_____	**14.**	10	_____	_____
9.	5	_____	_____	**15.**	11	_____	_____
10.	6	_____	_____	**16.**	12	_____	_____

You want to build a dog pen that encloses 36 ft².

17. What lengths, widths, and perimeters are possible? Round to the nearest hundredth.

18. The fence you wish to use for the dog pen costs $7.00 a foot. You can't spend more than $225 for the fence. What size pen can you construct? Give the cost of each size.

The perimeter of a rectangle is 24 ft. Find the possible lengths and widths. Then find the area.

	l	w	Area
19.	1	11	1(11) = 11 ft²
20.	2	_____	_____
21.	3	_____	_____
22.	4	_____	_____
23.	5	_____	_____
24.	6	_____	_____
25.	7	_____	_____
26.	8	_____	_____
27.	9	_____	_____
28.	10	_____	_____
29.	11	_____	_____
30.	12	_____	_____

You have 40 ft of fence to enclose a space for a dog pen.

31. List all the possible lengths, widths, and areas for a rectangular pen.

32. Use any shape you would like for the dog pen. Sketch each with the measurements of the sides.

33. Would any of the shapes you used in Exercise 6 give you a better dog pen for your dog than a rectangular pen? Why or why not?

Applications

A rectangular pool, 12 ft by 20 ft is situated in a yard, 50 ft by 30 ft.

34. Find the area of the pool and the yard.

35. What percent of the yard is covered by the pool?

36. Determine the size of a pool that would cover 25% of the yard.

A wallpaper border is to be purchased for a rectangular room 12 ft by 15 ft.

37. What is the perimeter of the room?

38. A roll of wallpaper border is 10 yds long. How many rolls will be needed?

39. If a roll costs $8.50, what is the cost of the border?

Name _____

Date _____

Square Roots and the Pythagorean Theorem

Objective You will be able to solve problems using squares, square roots, and the Pythagorean theorem.

Keys fx-7700G  [x^2]   **TI-81** 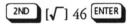 [$\sqrt{\ }$]

■ You can use a graphing calculator to find square roots by using the square root key. The display will show up to 10 digits in the display. You may want to round the square root to the nearest hundredth.

fx-7700G	TI-81

Example A Find $\sqrt{46}$.

[√] 46 [EXE]

Answer: 6.782329983 or 6.78 to the nearest hundredth

Find $\sqrt{46}$.

[2ND] [√] 46 [ENTER]

Answer: 6.782329983 or 6.78 to the nearest hundredth

The Pythagorean theorem is used to work with the lengths of the sides of a right triangle. The square of the length of the hypotenuse c is equal to the sum of the squares of the other two sides a and b, or $a^2 + b^2 = c^2$. To use a graphing calculator to find c, you can enter the entire calculation at once since $c = \sqrt{a^2 + b^2}$.

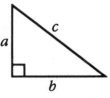

fx-7700G	TI-81

Example B Find c if $a = 7$ and $b = 5$.

Answer: 8.602325267 or 8.6 to the nearest hundredth.

Find c if $a = 7$ and $b = 5$.

Answer: 8.602325267 or 8.6 to the nearest hundredth.

Exercises

Find each square root to the nearest hundredth.

1. $\sqrt{6400}$ _____
2. $\sqrt{2116}$ _____
3. $\sqrt{59}$ _____
4. $\sqrt{4520}$ _____
5. $\sqrt{28.65}$ _____
6. $\sqrt{3.61}$ _____
7. $\sqrt{79.5664}$ _____
8. $\sqrt{9.765625}$ _____
9. $\sqrt{18.974}$ _____
10. $\sqrt{6760}$ _____
11. $\sqrt{9.216}$ _____
12. $\sqrt{0.216}$ _____

Find the missing sides. Round answers to the nearest hundredth.

13.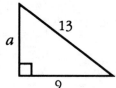

14.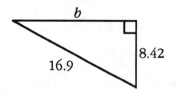

15.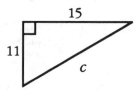

16.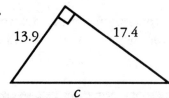

17.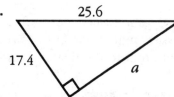

18.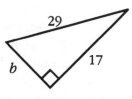

19.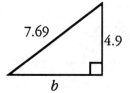

20.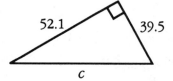

21.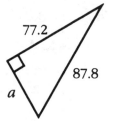

Drawing Conclusions

A rational number is a terminating decimal, such as $\sqrt{36} = 6$ and $\sqrt{1.742} = 1.32$.
An irrational number is a number which neither terminates nor repeats, such as $\sqrt{45}$.

22. For Exercises 1 - 12, identify each as a rational number or an irrational number.

An irrational number between 3 and 3.2 is $\sqrt{10}$ since $\sqrt{9} = 3$ and $\sqrt{10.24} = 3.2$.

23. Find four other irrational numbers between 3 and 3.2.

24. How many irrational numbers exist between 3 and 3.2? Why?

25. Find 5 irrational numbers between 7.2 and 7.3.

26. Examine your answers for Exercises 13 - 21. Describe the location of the longest side.

27. Explain why the diagram at the right is incorrect.

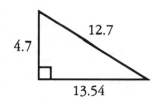

30

Applications

28. A 17-ft pole requires guy wires to stabilize it. Tom puts a stake 20 ft to the east of the pole. Allowing one foot on each end of the wire for tying, how long will the guy wire need to be?

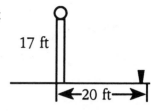

17 ft

←20 ft→

29. Jim puts another stake on the west side 15 ft out from the base. Again, allowing for wire to tie, how much wire is needed?

Challenge

A rectangular field is 1800 ft by 3000 ft. Tom and Travis are standing on the southwest corner. Travis and Tom walk at the same rate.

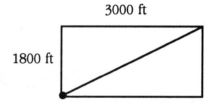

3000 ft

1800 ft

30. If Travis walks along the diagonal, straight to the opposite corner, how far does he walk?

31. If Tom walks along the edge of the field, straight east, then north, how far does he walk?

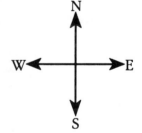

32. Who reaches the northeast corner first? Why?

33. If Tom walked twice as fast as Travis, will he beat Travis to the corner?

Name _____

Date _____

Area and Circumference
of a Circle

Objective   You will be able to find the area and circumference of a circle.

Keys   fx-7700G $\boxed{\rightarrow}$ TI-81 $\boxed{\text{STO} \blacktriangleright}$

■ A *circle* is the set of all points the same distance from a given point called the center. A *diameter* is a line segment that has endpoints on the circle and passes through the center of the circle. A *radius* is a line segment that has one endpoint on the circle and the other at the center.

The diameter d of a circle is twice its radius r. $d = 2r$

The circumference of a circle is the distance around the circle. The ratio of the circumference C to the diameter d is π. Store the approximation 3.14 for π in your calculator.

$\pi = \frac{C}{d}$
and
$C = \pi d$

The area A of a circle is the product of π and the square of the radius r. $A = \pi r^2$

fx-7700G	TI-81

Example A   A circle has radius 5. Find its diameter, circumference, and area.

A circle has radius 5. Find its diameter, circumference, and area.

Store 3.14 in P.
3.14 $\boxed{\rightarrow}$ $\boxed{\text{ALPHA}}$ [P] $\boxed{\text{EXE}}$

Store 3.14 in P.
3.14 $\boxed{\text{STO} \blacktriangleright}$ [P] $\boxed{\text{ENTER}}$

$d = 2r$ 2 $\boxed{\text{X}}$ 5 $\boxed{\text{EXE}}$
The diameter is 10 units.

2 $\boxed{\text{X}}$ 5 $\boxed{\text{ENTER}}$
The diameter is 10 units.

$C = \pi d$ $\boxed{\text{ALPHA}}$ [P] $\boxed{\text{X}}$ 10 $\boxed{\text{EXE}}$
The circumference is 31.4 units.

$\boxed{\text{ALPHA}}$ [P] $\boxed{\text{X}}$ 10 $\boxed{\text{ENTER}}$
The circumference is 31.4 units.

$A = \pi r^2$ $\boxed{\text{ALPHA}}$ [P] $\boxed{\text{X}}$ 5 $\boxed{\text{SHIFT}}$
[x^2] $\boxed{\text{EXE}}$
The area is 78.5 square units.

$\boxed{\text{ALPHA}}$ [P] $\boxed{\text{X}}$ 5 $\boxed{x^2}$ $\boxed{\text{ENTER}}$

The area is 78.5 square units.

Exercises

A circle has the given radius. Find its diameter, circumference, and area.

1. 7 in.   _____

2. 11 cm   _____

3. 15 ft   _____

4. 1.5 m   _____

Drawing Conclusions

Find 5 different circular objects. Using a tape measure, measure the length of each diameter and circumference. Find the ratio of C to d.

	Object	Diameter	Circumference	$\frac{C}{d}$
5.				
6.				
7.				
8.				
9.				

10. Examine the ratios in Exercises 5 - 9. The ratio $\frac{C}{d}$ should be π which is approximately 3.14. If some of your ratios are not close to 3.14, what might have caused the difference?

Applications

11. Tim rides his bike 1 mi to John's house. His bike has 26 in. diameter wheels. How many revolutions does his tire make during the trip to John's house? (1 mi = 5280 ft.)

12. Sonja has 48 ft of fence to make a pen. She can choose a square pen or a circular pen. Which encloses the most area?

Challenge

Carmen's family has a 16 ft circular pool in their backyard. The rectangular yard measures 20 ft by 40 ft.

13. Find the area covered by the pool. _____

14. Find the area covered by the yard. _____

15. What percent of the yard is covered by the pool? _____

16. Find what size circular pool would cover 35% of the backyard. _____

Name _____

Date _____

Graphing Equations

Objective You will be able to graph equations.

Keys **fx-7700G** GRAPH X,θ,T RANGE  **TI-81** Y= XIT GRAPH

[TRACE] = F1   [CLS] = F5   G↔T   RANGE TRACE

SHIFT [PRGM] F6 = :

■ In order to graph an equation with a graphing calculator, the equation must be in the form $y = mx + b$.

fx-7700G	TI-81
To enter an equation, press GRAPH.	To enter an equation, press Y=.

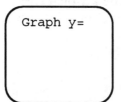

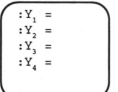

Input the equation using the key labeled X,θ,T for x.
Then press ENTER.
The graph will be displayed on the graph screen.

Input the equation using the key labeled XIT for x.
Then press GRAPH.
The graph will be displayed on the graph screen.

■ On both the *fx-7700G* and the *TI-81*, the RANGE key is used to set values for the portion of the graph that is displayed on the graph screen. The range values define a viewing rectangle. When you press the RANGE key, a menu will be displayed, as shown below.

```
Range
Xmin:-4.7    ←Minimum value on x-axis →
  max:4.7    ←Maximum value on x-axis →
  scl:1.     ←Scale used on x-axis     →
Ymin:-3.1    ←Minimum value on y-axis →
  max:3.1    ←Maximum value on y-axis →
  scl:1.     ←Scale used on y-axis     →
INIT
```

```
Range
Xmin=-4.8
Xmax=4.7
Xscl=1
Ymin=-3.1
Ymax=3.2
Yscl=1
Xres=1
```

On the *fx-7700G*, pressing F1, which is INIT, sets "friendly" range values of Xmin = –4.7, Xmax = 4.7, Xscl = 1, Ymin = –3.1, Ymax = 3.1, and Yscl = 1.

On the *TI-81*, to set "friendly" range values, enter values of Xmin = –4.8, Xmax = 4.7, Xscl = 1, Ymin = –3.1, Ymax = 3.2, Yscl = 1, and Xres = 1.

Graphing Equations

Example A

fx-7700G	**TI-81**
Graph $y = 3x - 4$.	Graph $y = 3x - 4$.

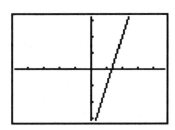

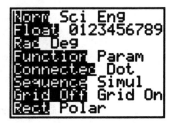

For the *TI-81*, if your graph does not look like the one shown check your mode setting. Press the [MODE] key. The mode screen should be highlighted as shown. If it is not, use the cursor keys and [ENTER] to highlight as shown. Then press [GRAPH].

■ Solutions to an equation can be found by using the trace feature to identify coordinates of points on the graph. After pressing [TRACE], you will see a blinking cursor on the screen. Use [▶] and [◀] to move the trace cursor along the graph. The *x*- and *y*-coordinates of the point shown by the blinking cursor are displayed at the bottom of the screen. The trace feature with the graph of Example A is shown below. For *fx-7700G*, the trace label is above the [F1] key.

fx-7700G	**TI-81**

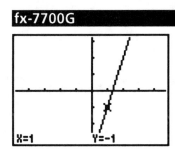

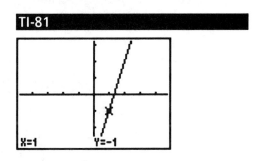

■ If the equation to be graphed is not in the form $y = mx + b$, then solve the equation for *y* before inputting it into the calculator.

Example B

fx-7700G

12*x* + 3*y* = 6.

First solve for *y*.
$$12x + 3y = 6$$
$$3y = -12x + 6$$
$$y = -4x + 2$$

[GRAPH] [−] 4 [X,θ,T] [+]
2 [EXE]

TI-81

Graph 12*x* + 3*y* = 6.

First solve for *y*.
$$12x + 3y = 6$$
$$3y = -12x + 6$$
$$y = -4x + 2$$

[Y=] [(−)] 4 [XIT] [+]
2 [GRAPH]

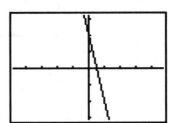

Note that you did not need to enter new range settings. However, if you changed the range settings, then reenter the range settings as in Example A.

■ The graph may be drawn over your previous graph. If you do not want this to happen, then clear the screen before entering a new graph. For the *fx-7700G*, use [CLS] [EXE]. For the *TI-81*, clear the screen by erasing any previous equations. Press [Y=], then place the cursor over any character in the equation you want to erase, and press [CLEAR]. To keep the equation, but not have a graph drawn, place the cursor on the equals symbol, and press [ENTER]. On the *fx-7700G*, to switch between the graph screen and the text screen, press the [G↔T] key. On the *TI-81*, to switch between the graph screen and the text screen use [CLEAR] and [GRAPH].

■ You can graph more than one equation at a time on a graphing calculator.

fx-7700G

Connect graphs by using the colon (:). The colon is found by pressing [SHIFT] [PRGM] [F6]. Note that the colon appears above [F6] in the program menu.

TI-81

When you press [Y=], there is space for four equations to be entered, y_1, y_2, y_3, and y_4.

LESSON 15   Graphing Equations

Example C

Graph $y = -3x + 4$ and $y = 2x - 6$.
Find the point of intersection.

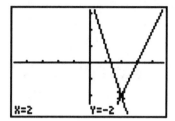

Graph $y = -3x + 4$ and $y = 2x - 6$.
Find the point of intersection.

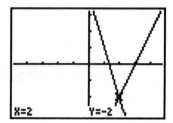

After the graph is drawn press [TRACE] which is $\boxed{F1}$. Use the cursor keys to find the point of intersection (2, -2). The $\boxed{\blacktriangle}$ and $\boxed{\blacktriangledown}$ keys move the cursor between the graphs.

After the graph is drawn press $\boxed{\text{TRACE}}$ Use the cursor keys to find the point of intersection (2, -2). The $\boxed{\blacktriangle}$ and $\boxed{\blacktriangledown}$ keys move the cursor between graphs.

Exercises

Graph each equation.

1. $y = -2x + 2$
2. $y = 3x + 5$
3. $y = 6x - 2$
4. $y = -5x + 2$
5. $y = -8x + 1$
6. $y = -8x - 1$
7. $y = 3x + 3$
8. $y = 4x - 3$
9. $y = -x + 3$

Solve for y and graph each equation.

10. $6x + 3y = 9$
11. $4x + 2y = -12$
12. $-24x + 8y = 12$
13. $-10x - 5y = 20$
14. $-28x + 7y = -14$
15. $30x - 9y = -18$

Drawing Conclusions

Graph. Then use the trace feature to find the points where the graphs intersect and where each graph intersects the x-axis and the y-axis.

16. $y = x + 2$ and $y = -2x - 1$
17. $y = -5x + 3$ and $y = 5x + 3$

Think about your answers to Exercises 16 and 17.

18. What is the value of y at the point where a graph crosses the x-axis?

19. What is the value of x at the point where a graph crosses the y-axis?

20. Determine two equations that contain the point (–1, 4). Graph your equations and use the trace feature to verify your answer. Could these graphs have another point in common?

Name _____

Date _____

Solving Equations by Graphing

Objective   You will be able to solve equations.

Keys   **fx-7700G** ⌊RANGE⌋   **TI-81** ⌊RANGE⌋

■ Recall that in solving an equation, you must keep the sides balanced. This means that each side has equal value. If you translate each side into a graph, the point where the two graphs cross is the point of balance or equality, which results in the solution to the equation. Remember to set the range as described in Lesson 15.

fx-7700G	**TI-81**

Example A   Solve $2x - 5 = 1$.   Solve $2x - 5 = 1$.

Graph $y = 2x - 5$ and $y = 1$.   Graph $y = 2x - 5$ and $y = 1$.

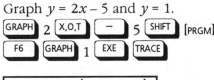

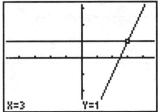   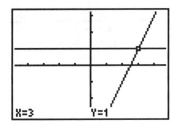

Since the point of intersection is (3, 1), $x = 3$ is the solution of the equation. Check by substituting 3 for x in the original equation.

$$2x - 5 = 1$$
$$2(3) - 5 \, ? \, 1$$
$$6 - 5 \, ? \, 1$$
$$1 = 1$$

■ Sometimes you may want to see more of the graph. You can do this by changing the range values. Choosing range values that depend on the screen size will give you a "friendly" viewing rectangle. To choose "friendly" values, you use the fact that the screen of the calculator is 96 dots wide and 64 dots deep.

fx-7700G	**TI-81**

The difference between Xmin and Xmax is a multiple of 94.

The difference between Xmin and Xmax is a multiple of 95.

$$\left.\begin{array}{l} \text{Xmin} = -4.7 \\ \text{Xmax} = 4.7 \end{array}\right\} 4.7 - (-4.7) = 9.4$$

$$\left.\begin{array}{l} \text{Xmin} = -4.8 \\ \text{Xmax} = 4.7 \end{array}\right\} 4.7 - (-4.8) = 9.5$$

LESSON 16   Solving Equations by Graphing

The difference between Ymin and
Ymax is a multiple of 62.

$$\left.\begin{array}{l} \text{Ymin} = -3.1 \\ \text{Ymax} = 3.1 \end{array}\right\} 3.1 - (-3.1) = 6.2$$

The difference between Ymin and
Ymax is a multiple of 63.

$$\left.\begin{array}{l} \text{Ymin} = -3.1 \\ \text{Ymax} = 3.2 \end{array}\right\} 3.2 - (-3.1) = 6.3$$

Some possible "friendly" windows are:

Xmin = –47	Xmin = –23.5	Xmin = –9.4	Xmin = –48	Xmin = –24	Xmin = –9.6
Xmin = 47	Xmin = 23.5	Xmin = 9.4	Xmin = 47	Xmin = 23.5	Xmin = 9.4
Ymin = –31	Ymin = –15.5	Ymin = –6.2	Ymin = –32	Ymin = –16	Ymin = –6.2
Ymin = 31	Ymin = 15.5	Ymin = 6.2	Ymin = 31	Ymin = 15.5	Ymin = 6.4

Exercises

Use the calculator to solve each equation for x. Choose a "friendly" window for your graph.
Check your results.

1. $x - 5 = -2$ _____

2. $9.6 + x = 12.3$ _____

3. $\frac{x}{3} = 3.2$ _____

4. $\frac{x}{2} + 12 = 10$ _____

5. $x + 3x - 1 = 11$ _____

6. $6x + 3 = 3x - 3$ _____

7. $4(3 - x) = 5(2x + 1)$ _____

8. $12(2x - 3) = 2(x + 4)$ _____

Drawing Conclusions

9. If Xmin is –940 and Ymin is –310, determine Xmax and Ymax for a
"friendly" window. What would be an appropriate scale?

_____

10. What is the significance of $y = 1$ in the solution to the example?
What does y represent?

_____

Applications

11. If a plane trip covered 1100 mi and took $2\frac{3}{4}$ h, approximate the cruising
speed of the plane.

12. If the Washington Monument were 85 ft shorter, its height would be as tall
as the Gateway Arch in St. Louis. If the height of the Gateway Arch were
480 ft less than twice the height of the Washington Monument, how tall is
each structure?

Name _____

The Slope of a Line

Date _____

Objective You will be able to determine the slope of a line from its graph.

Keys **fx-7700G** [PLOT] = [F3]   **TI-81** [2ND] [DRAW] [1] = ClrDraw

[LINE] = [F4]   [2ND] [DRAW] [2] = Line

[SHIFT] [,]   [ALPHA] [,]

■ The slope of a line is the ratio of its rise (vertical distance) compared to its run (horizontal distance).

slope = $\dfrac{\text{rise}}{\text{run}}$

To use a graphing calculator to find the slope of a line, first graph the equation. Then use the trace function to find two points on the line and determine the rise and run. Remember to use "friendly" windows for the range as in lesson 16.

fx-7700G	**TI-81**

Example A | Graph $y = -\frac{1}{2}x - 1$. Locate two points on the graph and determine the slope. | Graph $y = -\frac{1}{2}x - 1$. Locate two points on the graph and determine the slope. |

Locate (–2, 0) and (0, –1). The rise is –1 and the run is 2, so

slope = $\dfrac{\text{rise}}{\text{run}}$ = $-\dfrac{1}{2}$

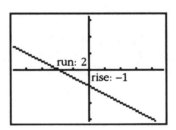

■ The graphing calculator can be used to plot two points and connect the points with a line segment as shown in Example B. The label [PLOT] is found above F3 and the label [LINE] is found above F4 on the *fx-7700G*. When plotting a point on the *fx-7700G*, you must press [EXE] twice. The label [DRAW] is found above [PRGM] on the *TI-81*.

LESSON 17 The Slope of a Line

fx-7700G

Example B Draw a line segment using the coordinates (–2, 1) and (1, –1). Then find the slope.

Move the cursor to $x = 1$, $y = –1$.

EXE [LINE] EXE

The slope is $-\dfrac{2}{3}$.

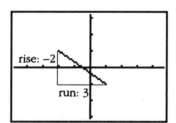

TI-81

Draw a line segment using the coordinates (–2, 1) and (1, –1). Then find the slope.

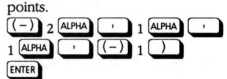

Enter the coordinates of the two points.

The slope is $-\dfrac{2}{3}$.

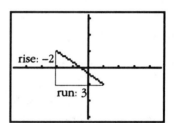

■ Clear the graph screen of your calculator as shown before each exercise.

Exercises

Graph each equation. Locate two points and determine the slope.

1. $y = \frac{2}{3}x + 1$ _____ **2.** $y = -\frac{1}{4}x - 2$ _____

3. $y = 2x - 2$ _____ **4.** $y = 0.1x + 2$ _____

5. $y = -4x + 1$ _____ **6.** $y = -2.5x - 2$ _____

7. $y = 0.25x - 3$ _____ **8.** $y = -\frac{2}{3}x - 6$ _____

9. $3x - 2y = 8$ _____ **10.** $6x + 3y = 9$ _____

Draw each line segment using the given coordinates. Then find the slope.

11. (0, 4) and (3, 5) _____ **12.** (–3, 2) and (1, 3) _____

13. (–2, 6) and (4, –8) _____ **14.** (2, –4) and (5, –1) _____

15. (–7, –3) and (–3, 2) _____ **16.** (–5, 0) and (4, –7) _____

Drawing Conclusions

Given the slope and the coordinates of one point, find another point on the line.

17. $m = \frac{2}{5}$ and (2, 6) _____   **18.** $m = -3$ and (–5, 3) _____

19. $m = -\frac{5}{4}$ and (1.3, –2.6) _____   **20.** $m = -\frac{3}{5}$ and (2, –3) _____

21. Rank the slopes found in Exercises 1 - 6 in order from least steep to steepest?

22. In Exercises 1 - 6, which slopes are positive? _____

Which slopes are negative? _____

Is there a relationship between the sign of a slope of a line and the direction the line slants from right to left? Explain.

Applications

23. Find some real life examples of how lines are used. Make a poster to share with the class. Use a graphing calculator to help you estimate equations for the lines.

Name

Date

Graphing Systems of Linear Equations

Objective   You will be able to solve systems of linear equations by graphing.

Keys   **fx-7700G** [TRACE]

[SHIFT] [PRGM] [F6]

TI-81 [TRACE]

■ Systems of linear equations can be solved by graphing. The point of intersection is the solution of the system since it is a solution of both equations. To use your graphing calculator, first graph both equations and then use the trace feature to determine the point of intersection.

Remember that on the *fx-7700G* you use the colon to connect graphs. To get the colon, press [SHIFT] [PRGM] [F6]. On the *TI-81*, simply enter one equation at y_1 and the other at y_2. Be sure to set the range to the basic friendly window.

fx-7700G	TI-81
Example A Solve this system of equations by graphing.	Solve this system of equations by graphing.

$y = \frac{1}{5}x - 1$
$y = -x + 2$

$y = \frac{1}{5}x - 1$
$y = -x + 2$

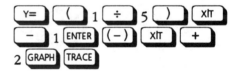

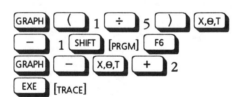

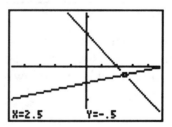

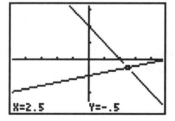

Move the cursor to the point of intersection and read the coordinates. The *x*-coordinate is 2.5 and the *y*-coordinate is –0.5. Check the solution in both equations.

$y = \frac{1}{5}x - 1$
$-0.5 \ ? \ \frac{1}{5}(2.5) - 1$
$-0.5 = -0.5$

$y = -x + 2$
$-0.5 \ ? \ -(2.5) + 2$
$-0.5 \ ? \ -0.5$

The solution is (2.5, –0.5).

■ Clear the graph screen before solving another system. On the *fx-7700G*, at the graph screen use [cls] $\boxed{\text{EXE}}$. On the *TI-81*, press $\boxed{\text{Y=}}$, then place the cursor over any character in the equation you want to erase and press $\boxed{\text{CLEAR}}$.

Remember that if an equation to be graphed is not in the form $y = mx + b$, then solve for y as in lesson 15 before you use the graphing calculator.

Exercises

Solve each system of equations by graphing.

1. $y = x - 2$
   $y = -\frac{1}{2}x + 1$ _____

2. $y = -2x + 8$
   $y = x - 4$ _____

3. $y = x$
   $y = -x + 3$ _____

4. $y = -2x + 1$
   $y = 3x - 4$ _____

5. $y = -3x + 6$
   $y = -4x + 7$ _____

6. $y = 3.5x - 0.5$
   $y = -0.5x + 1.5$ _____

7. $y = \frac{1}{3}x - 1$
   $y = 2x + 3$ _____

8. $y = 0.25x - 2$
   $y = -x + 2.5$ _____

9. $4y + 3x = 8$
   $y - x = -5$ _____

10. $4x + 5y = 12$
    $-4x + 3y = 4$ _____

11. $3y - 2x = 6$
    $y + 2x = 6$ _____

12. $x + 3y = 5$
    $y - x = 3$ _____

Drawing Conclusions

13. Alfie, an ant, starts on a grid at $(-2, 1)$ and walks in an L-shape, up 1 unit and then right 2 units. If he stops at $(6, 5)$, write an equation that represents the line Alfie should have followed to take the shortest path from $(-2, 1)$ to $(6, 5)$?

14. Consider that Alfie had a friend, Trey, moving on a straight line that would be represented by $y = -\frac{1}{2}x + 1$. Is it possible that Trey could meet up with Alfie? Where? Explain.

Applications

15. One kind of tea costs $1.69 per 3 oz package and a second kind costs $2.29 per 3 oz package. If Marcus spent a total of $16.52 for 8 packages, how many of each did he buy?

16. A group of 4 couples is going out for ice cream. If 4 people have ice cream sodas and 4 have sundaes, the bill will total $16.92. However, if only 2 people have sundaes and 6 have sodas, the bill will cost $16.38. What is the cost of each soda and sundae?

Challenge

A boat leaves a harbor located at (0, –1) and travels along a course given by $y = x - 1$. Another boat is at (4, 0) and travels along a course given by $y = -2x + 8$.

17. Determine where the boats will meet.

16. Which boat has further to travel?

Name _____

Date _____

**Graphing Inequalities:
A Two-Dimensional View**

Objective   You will be able to use the graphing calculator for determining inequality solutions on number lines.

Keys   **fx-7700G** [MODE] [SHIFT] [÷]
[MODE] [SHIFT] [+]

TI-81 [2ND] [DRAW] [7] (Shade)
[2ND] [DRAW] [1] (ClrDraw)
[ALPHA] [,]

■ In Lesson 16, you learned that each side of an equation can be represented by lines. The lines intersect at a value for *x* that gives the solution to the equation. A similar idea can be used for inequalities. Graph each side of the inequality. The point where the edges of the two graphs cross identifies the solution to the inequality. To graph an inequality on a graphing calculator, the inequality feature must be used. Remember to set "friendly" range values as described in Lesson 16.

■ For the *fx-7700G*, use [MODE] [SHIFT] [÷] for the inequality mode which creates shading (INEQ). When you press [GRAPH] a menu appears. Select [F1] (Y>), [F2] (Y<), [F3] (Y≥), or [F4] (Y≤). When graphing two sides of the inequality, be sure to enter the correct inequality.

For *x* < *b*, graph *y* > *x* and *y* < *b*.

For *x* > *b*, graph *y* < *x* and *y* > *b*.

Remember to use [SHIFT] [PRGM] [F6] (:) to draw two graphs. To return to equation mode, use [MODE] [SHIFT] [+] (REC). To clear the graph screen, use [SHIFT] [F5] (CLS) [EXE] .

■ For the *TI-81*, [2ND] [DRAW] [7] (Shade) allows you to enter expressions to be shaded. To use the shade command, you enter (first expression, second expression, *c*). The number *c* is used to set the resolution of the screen and can be an integer from 1 to 8.

If the inequality is *x* < *b*, then use Shade (*x*, *b*, *c*).

If the inequality is *x* > *b*, then use Shade (*b*, *x*, *c*).

Use [ALPHA] [,] for the comma. To clear the graph screen, use [2ND] [DRAW] [1] (ClrDraw) [ENTER] .

■ When drawing the graph of an inequality, remember to use a solid dot for ≤ and ≥, and an open dot for < and >.

LESSON 19 Graphing Inequalities: A Two-Dimensional View

fx-7700G

Example A Solve $x - 1 < 1$ and graph the solution on a number line.

Change to inequality mode.
[MODE] [SHIFT] [÷] (INEQ)

Clear the graph screen.
[SHIFT] [F5] (CLS) [EXE]

Set the range.
[RANGE] [F1] (INIT) [RANGE]
[RANGE]

Graph $y > x - 1$ and $y < 1$.
[GRAPH] [F1] (>) [X,θ,T] [–]
[1] [SHIFT] [PRGM] [F6] (:)
[GRAPH] [F2] (<) [1] [EXE]

Find the point of intersection.
[F1] (Trace)

The point of intersection is at $x = 2$.

Answer: $x < 2$

TI-81

Solve $x - 1 < 1$ and graph the solution on a number line.

Clear the graph screen.
[2ND] [DRAW] [1] (ClrDraw) [ENTER]

Set the range.

Xmin –4.8, Xmax 4.7, Xscl 1

Ymin –3.1, Ymax 3.2, Yscl 1

Graph $y > x$ and $y < 1$.
[2ND] [DRAW] [7] (Shade) [XT]
[–] [1] [ALPHA] [,] [1]
[ALPHA] [,] [3] [ )] [ENTER]

Find the point of intersection.
[TRACE]

The point of intersection is at $x = 2$.

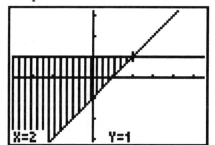

Exercises

Solve each inequality and graph the solution on a number line. Suggested range values are given for each calculator.

1. $x - 5 > 1$
   TI-81 (–0.8, 8.7, 1, –6.4, 6.2, 1)
   fx-7700G (-0.7, 8.7, 1, –6.2, 6.2, 1)

2. $-4x < 10$
   TI-81 (–4.8, 4.7, 1, –15.5, 16, 1)
   fx-7700G (–4.7, 4.7, 1, –15.5, 15.1, 1)

3. $x + 3 \le 8$
   TI-81 (–1.8, 7.7, 1, –0.4, 12.2, 1)
   fx-7700G (–1.8, 7.6, 1, –0.4, 12, 1)

4. $-x + 1 < -5$
   TI-81 (–1.8, 17.2, 1, –8.4, 4.2, 1)
   fx-7700G (–1.8, 17, 1, –8.2, 4.2, 1)

5. $3x - 1 < 11$
TI-81 (–1.8, 17.2, 1, –2.5, 16.4, 1)
fx-7700G (–1.8, 17, 1, –2.6, 15.9, 1)

6. $x + (-2x) \leq 3$
TI-81 (–4.8, 4.7, 1, –6.2, 6.4, 1)
fx-7700G (–4.7, 4.7, 1, –6.2, 6.2, 1)

7. $18 + 3x < x + 30$
TI-81 (–1.8, 7.7, 1, –4, 46.4, 1)
fx-7700G (–1.8, 7.6, 1, –4, 45.6, 1)

8. $2(x - 3) > 15 - x$
TI-81 (–1.8, 17.2, 1, –2, 10.6, 1)
fx-7700G (–1.8, 17, 1, –4, 45.6, 1)

Applications

Use the information at the right to set up inequalities and solve. Remember to choose an appropriate range.

Car Information
27 gallon tank
25 mi/gal city driving
35 mi/gal highway driving

9. If you started out with a full tank of gas and drove only in the city, draw a number line to represent how far you could probably go.

10. If you started out with a full tank of gas and drove only on the highway, draw a number line to represent how far you could probably go.

11. What are your miles per gallon if you did half city and half highway driving?

12. How far would you go on one tank using the answer you found to Exercise 11?

13. Explain how you determined the number in Exercise 12.

14. RESEARCH ▼
Investigate what is meant by one dimension, two dimensions, and three dimensions and prepare a report.

Name _____

Date _____

Objective   You will be able to find the average and draw a scatter plot for the given data.

Keys   fx-7700G MODE [,] PRE   TI-81 [STAT] [DRAW] VARS

■ You can enter data into your graphing calculator and draw a scatter plot. You can also find the average or mean, designated by the symbol $\overline{x}$ or $\overline{y}$.

$$\text{mean} = \frac{\text{sum of all the data}}{\text{the number of pieces of data}}$$

To use a graphing calculator, you need to enter the data using the correct menu.

■ For the *TI-81*, press 2ND [STAT] to get the statistics menu and press ▶ ▶ to get the data screen. Press 1 for Edit to enter data. To leave the data screen, press 2ND [QUIT] or select another screen such as 2ND [STAT]. To clear statistical data, use 2ND [STAT] ▶ ▶ (Data) 2 (ClrStat) ENTER.

To draw a scatter plot, first set a range that fits the data. Press 2ND [STAT] ▶ (Draw) to get the draw screen. Press 2 ENTER for a scatter plot. To clear a statistical drawing, press 2ND [DRAW] 1 (CLrDraw) ENTER.

■ For the *fx-7700G*, you need to set some special features in the mode menu. The mode display screen MDISP should appear as shown at the right.

MODE ÷ for regression mode (REG).
MODE 4 for linear regression (LIN).
MODE SHIFT 1 for storing statistical data (STO).
MODE SHIFT 3 for drawing a statistical graph (DRAW).
MODE SHIFT + for drawing a graph with rectangular coordinates (REC).
MODE SHIFT 5 for connecting the points plotted on the graph (CON).

```
    RUN/ LIN-REG
  S-data: STO
 S-graph: DRAW
  G-type: REC/CON
   angle: Deg
 display: Nrml
```

DT	EDIT	;	DEV	Σ	REG
F1	F2	F3	F4	F5	F6

Notice the labels for the function keys at the bottom of the screen. Use F1 (DT) to input data. Use F2 (Edit) to edit data.

If your screen looks like the screen at the right, press F6 G↔T to get the screen shown above. To clear statistical data, use F2 (EDIT) F3 (ERS = erase) F1 (YES).

DT	EDIT	;			CAL
F1	F2	F3	F4	F5	F6

To clear a statistical drawing, press [SHIFT] [F5] (Cls) [EXE]. If you clear the graph screen and your data is still in memory, press [MODE] 4 [F6] (CAL) to redraw the graph. To return to a previous menu, press [PRE], the previous key. Remember to use [G↔T] to move from the graph screen to the data screen.

■ Consider the data table at the right of the depth of the water in a reservoir. The months have been numbered so that the numbers can be entered into the calculator, since the names of each month cannot be entered. The data is entered into the calculator as two values so that the graphing options can be used.

	Month	Depth (meters)
1	November	4
2	December	3
3	January	5
4	February	6
5	March	5.5
6	April	7.5
7	May	8
8	June	9.5
9	July	6.5

Example A

fx-7700G

Make a scatter plot and find the average depth of the reservoir.

Clear the graph screen.
[SHIFT] [F5] (Cls) [EXE]
Set the mode as above.
Clear any previous statistical data.
[F2] (EDIT) [F3] (ERS) [F1] (YES)
Set the range.
Xmin 0, Xmax 10, Xscl 0.5
Ymin 0, Ymax 10, Yscl 0.5
Enter the data using the statistical menu. A dot will be placed on the graph screen as each data point is entered. To continue, just type the next number. Note that the comma is above the → key.
1 [SHIFT] [,] 4 [F1] (DT)
2 [SHIFT] [,] 3 [F1] (DT)
3 [SHIFT] [,] 5 [F1] (DT)
4 [SHIFT] [,] 6 [F1] (DT)
5 [SHIFT] [,] 5.5 [F1] (DT)
6 [SHIFT] [,] 7.5 [F1] (DT)
7 [SHIFT] [,] 8 [F1] (DT)
8 [SHIFT] [,] 9.5 [F1] (DT)
9 [SHIFT] [,] 6.5 [F1] (DT)

TI-81

Make a scatter plot and find the average depth of the reservoir.

Clear the graph screen.
[Y=] [CLEAR] for each equation.
[2ND] [DRAW] [1] (CLrDraw) [ENTER]
Clear any previous statistical data.
[2ND] [STAT] [▶] [▶] (Data)
[2] (ClrStat) [ENTER]
Set the range.
Xmin 0, Xmax 10, Xscl 0.5
Ymin 0, Ymax 10, Yscl 0.5
Enter the data.
[2ND] [STAT] [▶] [▶] (Data)
[1] (Edit)
1 [ENTER] 4 [ENTER]
2 [ENTER] 3 [ENTER]
3 [ENTER] 5 [ENTER]
4 [ENTER] 6 [ENTER]
5 [ENTER] 5.5 [ENTER]
6 [ENTER] 7.5 [ENTER]
7 [ENTER] 8 [ENTER]
8 [ENTER] 9.5 [ENTER]
9 [ENTER] 6.5 [ENTER]
Draw a scatter plot.
[2ND] [STAT] [▶] (Draw) [2ND]
(Scatter) [ENTER]

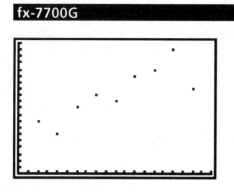

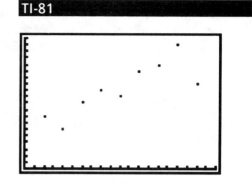

Find the average.

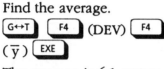

($\overline{y}$) [EXE]

The average is 6.1 meters.

Find the average.

[2ND] [stat] [2] (LinReg) [ENTER]

[VARS] [5] ($\overline{y}$) [ENTER]

The average is 6.1 meters.

Exercises

Make a scatter plot and find the average or mean for each set of data.

1. A telephone company is studying its long-distance charges. The length in minutes of 30 calls are given below. Enter the data numbers as *x*-values and enter the data points as *y*-values. Use these range values: Xmin 0, Xmax 30, Xscl 1, Ymin 0, Ymax 40, Yscl 1.

22	17	5	11	31	13	6	10	23	14	9	5	18	37	15
12	12	30	19	14	10	8	26	40	27	21	9	6	51	48

2. The population of the Earth in millions is given below. Use these range values: Xmin 1600, Xmax 1990, Xscl 50, Ymin 0, Ymax 5500, Yscl 500.

x	1650	1700	1750	1800	1850	1900	1950	1990
y	550	600	725	900	1200	1500	4340	5200

3. In a survey on doing sit-ups, 50 students participated. Each student completed as many sit-ups as possible as shown below. Enter the data numbers as *x*-values and enter the data points as *y*-values. Use these range values: Xmin 0, Xmax 50, Xscl 1, Ymin 0, Ymax 80, Yscl 1.

51	43	76	26	63	49	12	32	59	42	48	45	34	27	12	91	56
57	13	24	66	37	27	45	53	47	65	54	15	18	56	55	83	35
75	18	39	72	10	23	28	52	36	67	46	58	38	48	68	51	

4. The age of each United States president at inauguration is given below. Enter the data numbers as *x*-values and enter the data points as *y*-values. Use these range values: Xmin 0, Xmax 42, Xscl 1, Ymin 35, Ymax 70, Yscl 1.

57	61	57	57	58	57	61	54	68	51	49	64	50	48	65
52	56	46	54	49	50	47	55	55	54	42	51	56	55	51
54	51	60	62	43	55	56	61	52	69	64	46			

5. The time in seconds to run 100 meters was recorded for some students. Enter the data numbers as *x*-values and enter the data points as *y*-values. Use these range values: Xmin 0, Xmax 40, Xscl 1, Ymin 12.5, Ymax 18.5, Yscl 0.5.

16.2	17.6	15.0	15.6	15.3	13.1	17.1	16.2	15.6	14.5
14.1	18.0	13.1	17.2	18.0	16.8	15.1	13.1	16.8	18.1
15.2	15.5	16.6	13.7	15.5	14.7	17.9	16.0	18.1	15.3
15.7	17.4	14.1	16.4	15.3	16.6	13.0	17.3	14.7	15.5

Drawing Conclusions

The table gives the number of students who bought lunch in the cafeteria each day for four weeks.

	Mon	Tue	Wed	Thu	Fri
Week 1	154	146	192	139	141
Week 2	149	151	181	132	151
Week 3	138	142	190	146	158
Week 4	143	149	184	151	147

6. What was the average for each day of the week?

7. What was the average for each week?

8. What was the overall average for the month?

9. Compare the different averages. Discuss how they relate. How would the averages help the cafeteria staff with planning menus?

Applications

10. The monthly normal temperatures in Atlanta, Georgia are as follows:

Jan 42	Feb 45	Mar 53	Apr 62	May 69	Jun 76
Jul 79	Aug 78	Sep 73	Oct 62	Nov 52	Dec 45

    Make a scatter plot of the data using an appropriate range. Is it appropriate to discuss the average temperature for Atlanta? Why? Why not?

11. The table gives the number of passes attempted and the number of completions by a quarterback for each game of a season. What kind of information can you provide for the team manager if you use averages with the game data?

Game	1	2	3	4	5	6	7	8	9
Attempts	26	32	27	23	28	20	28	20	25
Completions	17	10	15	8	12	10	18	11	6

12. **Research ▼**
    Collect temperature data or call the weather bureau in your area to obtain temperature data for a given month. What predictions or conclusions can you draw from calculating averages?

Name _____

Exploring Data and
Line Relationships

Date _____

Objective You will be able to find the line that best fits a given set of data.

Keys fx-7700G [GRAPH] [SHIFT] [F4] (LIN)   TI-81 [stat] [draw] [VARS]

■ You may have noticed that when you make a scatter plot of data, sometimes it looks like all the points are close to some line. You can use the graphing calculator to find the line that best fits a set of data. Remember that the equation of a line is of the form $y = a + bx$. You can use a graphing calculator to automatically calculate a, the y-intercept, and b, the slope of the line. The calculator also determines a value for r, the correlation coefficient, which is a measure of how good the fit is. If the value of the correlation coefficient r is close to +1 or –1, the fit is very good. A correlation coefficient of 0.79 is considered to be a moderately good fit.

fx-7700G	TI-81

Example A Find the line of best fit for the given data set.

Find the line of best fit for the given data set.

x	1	2	3	4	5	6	7	8	9	10	11	12	13	14	15
y	3.8	3.6	3.5	3.2	3.2	3	2.8	2.8	2.8	2.4	2.2	2.1	2	2	2.1

Clear the graph screen.

[SHIFT] [F5] (Cls) [EXE]

Set the mode as in Lesson 20.

Clear any previous statistical data.

[F2] (EDIT) [F3] (ERS)

[F1] (YES).

Set the range.

Xmin 0, Xmax 15, Xscl 1

Ymin 0, Ymax 4, Yscl 0.1

Then enter the data using the statistical menu. A dot will be placed on the graph screen as each data point is entered. To continue, just type the next number.

1 [SHIFT] [,] 3.8 [F1] (DT)

2 [SHIFT] [,] 3.6 [F1] (DT)

3 [SHIFT] [,] 3.5 [F1] (DT)

4 [SHIFT] [,] 3.2 [F1] (DT)

5 [SHIFT] [,] 3.2 [F1] (DT)

6 [SHIFT] [,] 3 [F1] (DT)

7 [SHIFT] [,] 2.8 [F1] (DT)

Clear the graph screen.

[Y=] [CLEAR] for each equation.

[2ND] [DRAW] [1] (ClrDraw) [ENTER]

Clear any previous statistical data.

[2ND] [STAT] [▶] [▶] (Data)

[2] (ClrStat) [ENTER]

Set the range.

Xmin 0, Xmax 15, Xscl 1

Ymin 0, Ymax 4, Yscl 0.1

Enter the data.

[2ND] [STAT] [▶] [▶] (Data)

[1] (Edit)

1 [ENTER] 3.8 [ENTER]

2 [ENTER] 3.6 [ENTER]

3 [ENTER] 3.5 [ENTER]

4 [ENTER] 3.2 [ENTER]

5 [ENTER] 3.2 [ENTER]

6 [ENTER] 3 [ENTER]

7 [ENTER] 2.8 [ENTER]

8 [ENTER] 2.8 [ENTER]

9 [ENTER] 2.8 [ENTER]

fx-7700G

8 [SHIFT] [,] 2.8 [F1] (DT)
9 [SHIFT] [,] 2.8 [F1] (DT)
10 [SHIFT] [,] 2.4 [F1] (DT)
11 [SHIFT] [,] 2.2 [F1] (DT)
12 [SHIFT] [,] 2.1 [F1] (DT)
13 [SHIFT] [,] 2 [F1] (DT)
14 [SHIFT] [,] 2 [F1] (DT)
15 [SHIFT] [,] 2.1 [F1] (DT)

Calculate *a*, *b*, and *r* for the line of best fit.

[G↔T]
[F6] (REG) [F1] (A) [EXE]
[F2] (B) [EXE] [F3] (r) [EXE]

a = 3.840952381
b = –.1342857143
r = –.980373018

For this data, the line of best fit is *y* = 3.84 – 0.13*x*. It is a fairly good fit since *r* = –0.98, which is close to –1.

Graph the line of best fit.

[GRAPH] [SHIFT] [F4] (Line)
[1] [EXE]

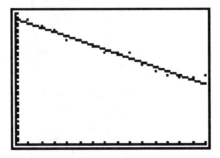

TI-81

10 [ENTER] 2.4 [ENTER]
11 [ENTER] 2.2 [ENTER]
12 [ENTER] 2.1 [ENTER]
13 [ENTER] 2 [ENTER]
14 [ENTER] 2 [ENTER]
15 [ENTER] 2.1 [ENTER]

Calculate *a*, *b*, and *r* for the line of best fit.

[2ND] [STAT] [2] (LinReg) [ENTER]

a = 3.840952381
b = –.1342857143
r = –.9803730181

For this data, the line of best fit is *y* = 3.84 – 0.13*x*. It is a fairly good fit since *r* = –0.98, which is close to –1.

Graph the line of best fit.

[Y=] [VARS] [▶] [▶] (LR)
[4] (RegEQ) [GRAPH]

Draw the scatter plot.

[2ND] [STAT] [▶] (Draw) [2]
(Scatter) [ENTER]

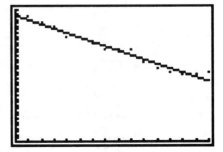

Exercises

Find the line of best fit for each set of data. Use appropriate range values.

1. | x | 19 | 19.5 | 20 | 20.5 | 21 | 22 | 22.5 | 23 | 23.5 | 24 |
   |---|---|---|---|---|---|---|---|---|---|---|
   | y | 10.5 | 10.3 | 10.25 | 10.2 | 10.1 | 9.85 | 9.8 | 9.79 | 9.7 | 9.6 |

2. | x | 88 | 99 | 110 | 121 | 132 | 143 | 154 | 165 | 176 | 187 | 198 | 209 |
   |---|---|---|---|---|---|---|---|---|---|---|---|---|
   | y | 60 | 65 | 70 | 75 | 80 | 85 | 90 | 95 | 100 | 105 | 110 | 115 |

3. | x | 66 | 75 | 83 | 91 | 100 | 108 | 115 | 125 | 133 | 141 | 150 | 158 |
   |---|---|---|---|---|---|---|---|---|---|---|---|---|
   | y | 88 | 99 | 110 | 121 | 132 | 143 | 154 | 165 | 176 | 187 | 198 | 209 |

4. | x | 24 | 28 | 32 | 36 | 48 | 52 | 56 | 60 | 62 | 68 | 72 | 76 |
   |---|---|---|---|---|---|---|---|---|---|---|---|---|
   | y | 6.02 | 5.42 | 5.28 | 5.26 | 5.17 | 5.12 | 4.54 | 4.5 | 4.43 | 4.31 | 4.19 | 4.09 |

Drawing Conclusions

5. Discuss how good a fit each line is for the data set of Exercises 1 - 4.

Applications

6. It is said that the measurement from the elbow crease on the inside of the right arm to the right wrist crease is the same length as that of the right foot. If this is true, then the line of best fit should have a slope of 1. Enter the following arm and foot data. Find the line of best fit and decide whether the lengths are equal.

arm (cm)	23	24	24	25	27	23	21	24	25	23.5	24	26.5
foot (cm)	23	23.5	24	24	27	24	23	24	25	24	24	26

7. **RESEARCH ▼**

   Decide on another relationship that appears to be linear. Collect data and prove or disprove your assumption by finding the line of best fit.

Lesson 2 Order of Operations

1. Divide 42 by 14. Multiply quotient by 12. Divide product by 9. Result is 4.
2. Work inside the parentheses. (Difference of 8 and 4 is 4. Difference of 15 and 8 is 7.) Multiply 3 by 4 and 6 by 7. Divide 42 by 3. Add 12 to the quotient 14. Result is 26.
3. Work inside the parentheses. (Subtract 8 from 15.) Multiply 3 by 8 and 6 by 7. Add products 24 and 42. Result is 66.
4. Work inside the parentheses. (Subtract 3 from 9). Divide 18 by 6. Multiply quotient by 2. Result is 6. Subtract 5 from 29. Divide difference by 6. The result is 4.
5. 8
6. 3
7. 51
8. 15
9. –6
10. 3
11. –21
12. 39
13. 195
14. 237
15. 56
16. 191
17. $16 \div 4 + 2 \times 8 = 20$
18. $16 \div (4 + 2) \times 8 = 21\frac{1}{3}$
19. $16 \div ((4 + 2) \times 8) = \frac{1}{3}$
20. $16 \div (4 + 2 \times 8) = 0.8$
21. $(16 \div 4 + 2) \times 8 = 48$
22. Whatever comes first from left to right.
23. Whatever comes first from left to right.
24. multiplication
25. division
26. 1 addition in parentheses; 2 multiplication; 3 subtraction
27. Answers may vary. Examples:
$1 = (9 \times 1) – (4 \times 2)$
$2 = 9 – (4 + 2 + 1)$
$3 = (1 + 9 + 2) \div 4$
$4 = (9 – 1) \div 4 + 2$
$5 = (9 – 4) \times (2 – 1)$
$6 = (9 + 2) – (1 + 4)$
$7 = 9 \div (1 + 2) + 4$
$8 = 1 + 9 + 2 – 4$
$9 = 9 \div (4 – 2 – 1)$
$10 = 9 + (4 – 2 – 1)$

Lesson 3 Adding Integers

1. 25
2. 1
3. –9
4. –26
5. 6
6. 11
7. 1
8. 18
9. 6
10. –11
11. 7
12. 4
13. 0
14. –32
15. 1
16. –1
17. 0
18. 15
19. positive
20. negative
21. The sign of the number farther from zero, or zero.
22. zero
23.

2	–5	0
7	1	–39
11	–1	–4

24.

4	–5	15
6	0	–5
7	–1	–20

25. $8 + (-6) + 7 = 9$; $30 + 9 = 39$; You have only gained 9 yards so you must kick from the 39 yard line.
26. 67°; $68 + (–4) + (–2) + 0 + 5 = 67$
27. Answers may vary.

Lesson 4 Subtracting Integers

1. –1
2. –25
3. –3
4. 8
5. –32
6. 3
7. –12
8. –46
9. –20
10. –5
11. 64
12. –196
13. 9
14. 0
15. –401
16. 12,001
17. –1
18. 23
19. zero
20. –2; –2; 4; 4; 9; 9; –5; –5; Answers may vary. Adding a negative integer is the same as subtracting a positive integer. Subtracting a negative integer is the same as adding a positive integer.
21. Overdrawn: $4.33; $123.41 – 25.76 – 36.75 – 65.23 = –4.33$
22. 168°F; $108 – (–60) = 168$
23. Answers may vary.
24. Magic square with sum 15.

25. Magic square with sum –6.

9	–6	–9
–20	–2	16
5	2	–13

Lesson 5 Multiplying Integers

1. 30
2. –14
3. –238
4. 483
5. –368
6. 316,224
7. –3451
8. –4928
9. –150,075
10. –54,756
11. 29,929
12. –50,010
13. 884
14. –1340
15. –125
16. 1
17. positive
18. negative
19. positive
20. Answer positive, if even number of negative integers multiplied. Answer negative, if odd number of negative integers multiplied.
21. Multiply –6 by 7.
22.

36	–39	15
–42	27	0
48	–72	27

23. –48°F; $2 \times (–24) = –48$
24. Total change in price: down $6.25; $5 \times (–1.25) = –6.25$; New stock price: $32.75: $39 – 6.25 = \$32.75$

Lesson 6 Dividing Integers

1. 58
2. 33
3. –309
4. –16
5. –12,500
6. 51
7. –20
8. 91
9. 6
10. –18
11. positive
12. positive
13. negative
14. A list of divided integers has a negative result if there are an odd number of negative integers and a positive result if there are an even number of negative integers.
15. 175
16. –14
17. –36
18. Yes. Answers may vary.
19. Each student owes $12.00.

Answers

20. Each student owes $13.64.
21. For exercise 19, each student owes $7.00. For exercise 20, each student owes $7.96. 300 − 125 = 175; 175 ÷ 25 = 7; 175 ÷ 22 = 7.9545
22. 5 stops are needed; −5290 ÷ 1000 = −5.29
23. The average is −21; −92 + (−75) + 32 + (−17) + 47 = −105; −105 ÷ 5 = −21
24. Answers may vary. If one is positive, then 2 must be positive and 3 negative, since an odd number must be negative. Example: 40, −32, −65, −43, 25

Lesson 7 Exponents and Integers

1. −16,384   2. 262,144
3. 48,828,125 4. 1
5. 20,736   6. 60,466,176
7. 43,046,721 8. 5,764,801
9. −2,097,152 10. 16,777,216
11. −512   12. 1024
13. 729   14. −2187
15. −52   16. −2
17. 128   18. 92
19. −128   20. 46
21. 32   22. 80
23. positive   24. negative
25. −125; −125; −125
    −15,625; 15,625; −15,625
    Answers may vary. Parentheses are needed to distinguish between raising a negative number to a power and the negative of a positive number raised to a power.
26. $800
27. $8,000
28. $80,000
29. Yes. The number of zeros in the interest is one less than the number of zeros in the principal amount. $800,000.
30. Across   Down
    1. −50   1. −51
    2. −70   2. −713
    3. −265   3. −27
    6. 10,117   4. 6605
    8. −1764   5. 54
    9. 13   7. 11

10. −22   8. −12
11. 0   10. −26
12. −80   12. −87
13. −36   13. −30
14. 256   14. 20
16. 7   15. 614
17. 630   17. 68
19. −50   18. 3031
20. −180   19. −51
21. 81   20. −10
22. −54   21. 83
23. 10   22. −55
24. 343   23. 16
26. −65   24. 31
27. 219   25. 493
28. −11   26. −61
29. −3   27. 225
30. −12   28. −13
31. 353   29. −380
33. 18   30. −16
34. 0   32. 51

Lesson 8 Fraction and Decimal Patterns

1. 0.5   2. 0.375
3. 0.4
4. 0.6666666667 or $0.\overline{6}$
5. 1.25   6. 0.125
7. 0.8333333333 or $0.8\overline{3}$
8. 0.4545454545 or $0.\overline{45}$
9. 1.571428571 or $1.\overline{571428}$
10. 0.2
11. 0.4285714286 or $0.\overline{428571}$
12. 0.3703703704 or $0.\overline{370}$
13. 0.1111111111 or $0.\overline{1}$
14. 0.2222222222 or $0.\overline{2}$
15. 0.3333333333 or $0.\overline{3}$
16. 0.1313131313 or $0.\overline{13}$
17. 0.3737373737 or $0.\overline{37}$
18. 0.6767676767 or $0.\overline{67}$
19. When dividing by nine the decimal repeats the numerator of the fraction.
20. $1.\overline{428571}$ 3   1 3/7
21. $11.\overline{1}$ 1   11 1/9
22. $80.\overline{3}$ 1   80 1/3
23. $51.\overline{2}$ 1   51 1/5
24. $21.\overline{375}$ 3   21 3/8
25. $0.\overline{285714}$ 26. $0.\overline{571428}$
27. $0.\overline{714285}$ 28. $0.\overline{857142}$
29. The digits that repeat are in the same order: 2, 8, 5, 7, 1, and 4. Each one starts off with a different first number.
30. Answers may vary.

Lesson 9 Fraction and Decimal Operations

Across		Down	
1.	210	1.	24
3.	154	2.	0.053
6.	34	3.	1
7.	0.525	4.	52
9.	231	5.	432
11.	−23	6.	399
12.	18	8.	0.51
13.	23	10.	132
14.	913	11.	−233
16.	1	15.	105.5
17.	0.03	16.	0.1534
18.	5/8	19.	0.625
19.	650	20.	0.0779
21.	412	22.	0.1327
24.	76	23.	0.25
25.	1	25.	14
27.	507	28.	01
29.	34	30.	30
30.	32		
31.	192		
32.	0.007		

Lesson 10 Evaluating Variable Expressions

1. −10; 2; 5; 17; 38; 54.5
2. 55; 51.4; 25; −8; −59; −95
3. 81; 75.42; 9; −9; −63; −135
4. 9   5. −11
6. 2.5   7. −5
8. −14.1   9. −2
10. 1.4   11. 8.1
12. 24°C   13. 38°C
14. 10°C   15. 0°C
16. 61°C   17. 10.5
18. 125   19. 42
20. 4   21. 24
22. 106.4528698
23. $1722.\overline{18}$ 24. −118.7

Lesson 11 Investigating Percents

1. $\frac{1}{4}$, $\frac{25}{100}$, $\frac{11}{44}$, $\frac{5}{20}$, $\frac{13}{52}$, $\frac{2}{8}$, $\frac{21}{84}$, $\frac{24}{96}$

1	25	11	27	6	
5	3	24	34		9
13	2	21		18	196
32	15		100	4	20
9		30	84	81	8
	18	68	96	52	44

Answers

2. percent symbol, %

3.

4. 40%   **5.** 10 years

6. Annual: 16 years; Quarterly: 15 years

7. Compounded annually: doubles in 10 years, triples in 15 years. Compounded quarterly: doubles in 9 years, triples in 14 years.

8. $66.06; 117.95 − 0.30(117.95) = 82.57; 82.57 − 0.20(82.57) = 66.06

9. Yes, then the coat would have cost $58.98; 117.95 − 0.50(117.95) = 58.98

10. 22.4 oz   **11.** 60%

12. 10% increase is wanted to go from 60% to 70%.

13. Answers may vary.

Lesson 12 Perimeter and Area Investigation

1. $P = 23.6$ cm; $A = 28.56$ cm²
2. $P = 20$ in; $A = 18.75$ in²
3. $P = 34$ ft; $A = 67.56$ ft²
4. $P = 67.4$ m; $A = 278.4$ m²
6. $w = 6$; $P = 16$
7. $w = 4$; $P = 14$
8. $w = 3$; $P = 14$
9. $w = 2.4$; $P = 14.8$
10. $w = 2$; $P = 16$
11. $w = 1.71$; $P = 17.42$
12. $w = 1.5$; $P = 19$
13. $w = 1.33$; $P = 20.67$
14. $w = 1.2$; $P = 22.4$
15. $w = 1.09$; $P = 24.18$
16. $w = 1$; $P = 26$
17.

l	w	P
1	36	74
2	18	40
3	12	30
4	9	26
5	7.2	24.4
6	6	24
7	5.14	24.28
8	4.5	25
10	3.6	27.2
11	3.27	28.54
13	2.77	31.54
14	2.57	33.14
15	2.4	34.8
16	2.25	36.5
17	2.12	38.24
19	1.89	41.78
20	1.8	43.6
21	1.71	45.42
22	1.64	47.28
23	1.57	49.14
24	1.5	51
25	1.44	52.88
26	1.38	54.76
27	1.33	56.66
28	1.29	58.58
29	1.24	60.40
30	1.2	62.4
31	1.16	64.32
32	1.13	66.26
33	1.09	68.18
34	1.06	70.12
35	1.03	72.06

18. You can use lengths of 3 ft to 13 ft. 3 by 12: $210; 4 by 9: $182; 5 by 7.2: $170.80; 6 by 6: $168; 7 by 5.14: $169.96; 8 by 4.5: $175; 9 by 4: $182; 10 by 3.6: $190.40; 11 by 3.7: $199.78; 12 by 3: $210; 13 by 2.77: $220.78

20. $w = 10$; $A = 20$ ft²
21. $w = 9$; $A = 27$ ft²
22. $w = 8$; $A = 32$ ft²
23. $w = 7$; $A = 35$ ft²
24. $w = 6$; $A = 36$ ft²
25. $w = 5$; $A = 35$ ft²
26. $w = 4$; $A = 32$ ft²
27. $w = 3$; $A = 27$ ft²
28. $w = 2$; $A = 20$ ft²
29. $w = 1$; $A = 11$ ft²
30. not possible
31.

l	w	A
1	19	19
2	18	36
3	17	51
4	16	64
5	15	75
6	14	84
7	13	91
8	12	96
9	11	99
10	10	100

32. Answers may vary. Example: a pentagon with side 8, an equilateral triangle with side 13.5, a hexagon with side $6\frac{2}{3}$, a trapezoid with bases 12 and 20 and side 4.

33. Answers may vary. A regular polygon that approximates a circle has the most area.

34. Area of the pool: 240 ft²
Area of the yard: 1500 ft²

35. 16%; 240 ÷ 1500 = 0.16

36. A pool with an area of 375 ft² will cover 25% of the yard; 1500 x 0.25. Answers may vary. Possible dimensions are 15 ft by 25 ft.

37. 54 ft

38. 2 rolls; 54 ÷ 30 ft = 1.8

39. $17; 2 x 8.50

Lesson 13 Square Roots and the Pythagorean Theorem

1. 80   **2.** 46
3. 7.68   **4.** 67.23
5. 5.35   **6.** 1.9
7. 8.92   **8.** 3.13
9. 4.36   **10.** 82.22
11. 3.04   **12.** 0.46
13. 9.38   **14.** 14.65
15. 18.6   **16.** 22.27
17. 18.78   **18.** 23.49
19. 5.93   **20.** 65.38
21. 41.82
22.
  1. rat.   **2.** rat.
  3. irrat.   **4.** irrat.
  5. irrat.   **6.** rat.
  7. rat.   **8.** rat.
  9. irrat.   **10.** irrat.
  11. irrat.   **12.** irrat.

23. Answers may vary. Example: $\sqrt{10.1}$, $\sqrt{10.2}$, $\sqrt{10.11}$, $\sqrt{10.15}$

24. Infinitely many because you can always find another number.

25. Answers may vary. For example, $\sqrt{53}$, $\sqrt{52}$, $\sqrt{52.5}$, $\sqrt{52.8}$, $\sqrt{52.25}$

26. The longest side is always located opposite the right angle.

27. The longest side is not opposite the right angle.

Answers

28. 28 ft; $2+\sqrt{17^2+20^2}=28.25$

29. 25 ft; $2+\sqrt{17^2+15^2}=24.67$

30. 3499 ft; $\sqrt{1800^2+3000^2}=$ 3498.57

31. 4800 ft; $3000+1800=4800$

32. Travis because it is the shorter distance.

33. Yes; $4800 \div 2 = 2400$ which is less than 3499

Lesson 14 Area and Circumference of a Circle

1. $d = 14$ in.; $C = 43.96$ in.; $A = 153.86$ in.²

2. $d = 22$ cm; $C = 69.08$ cm; $A = 379.94$ cm²

3. $d = 30$ ft; $C = 94.2$ ft; $A = 706.5$ ft²

4. $d = 3$ m; $C = 9.42$ m; $A = 7.065$ m²

5. – 9. Answers may vary.

10. Not measuring accurately, estimating incorrectly.

11. 776 revolutions; $C = 26$ x $3.14 = 81.64$; 5280 x $12 \div 81.64 = 776.09$

12. A circular pen encloses the most area (183.4 ft² for the circular pen and 144 ft² for the square pen).

13. 200.96 ft²

14. 800 ft²

15. Approximately 25%

16. A circular pool that has a diameter of about 18.9 ft.

Lesson 15 Graphing Equations

1.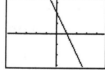

2.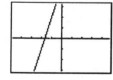

3.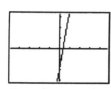

4.

5.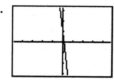

6.

7.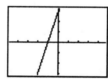

8.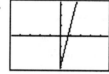

9.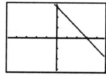

10. $y = -2x + 3$
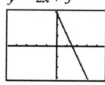

11. $y = -2x - 6$
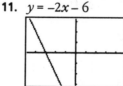

12. $y = 3x + \frac{3}{2}$
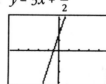

13. $y = -2x - 4$
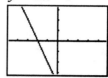

14. $y = 4x - 2$
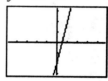

15. $y = \frac{10}{3}x + 2$
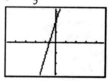

16.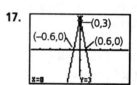

17.

18. zero   **19.** zero

20. Answers may vary. Example: $y = x + 5$ and $y = 2x + 6$; No other point in common unless the same line. Example: $y = x + 5$ and $2y = 2x + 10$.

Lesson 16 Solving Equations by Graphing

1. $x = 3$ **2.** $x = 2.7$

3. $x = 9.6$ **4.** $x = -4$

5. $x = 3$ **6.** $x = -2$

7. $x = 0.5$ or $\frac{1}{2}$ **8.** $x = 2$

9. For *TI-81*:   For *fx-7700G*:
Xmax 960   Xmax 940
Ymax 320   Ymax 310
Scale 100   Scale 100

10. The y value represents the value of the right or left expression when x is replaced by its solution.

11. 400 miles per hour.

12. Gateway Arch: 310 ft
Washington Mon.: 395 ft
$x - 85 = y$, $y = 2x - 480$

Answers

Lesson 17 The Slope of a Line

1. $\frac{2}{3}$ 2. $-\frac{1}{4}$ 3. 2

4. $\frac{1}{10}$ 5. -4

6. $-\frac{5}{2}$ or -2.5 7. $\frac{1}{4}$

8. $-\frac{2}{3}$ 9. $\frac{3}{2}$ 10. -2

11. $\frac{1}{3}$ 12. $\frac{1}{4}$ 13. $-\frac{7}{3}$

14. 1 15. $\frac{5}{4}$ 16. $-\frac{7}{9}$

17. – 20. Answers may vary. Examples are shown.

17. (7, 8) 18. (–4, 0)

19. (5.3, –7.6) 20. (7, –6)

21. $\frac{1}{10}$, $-\frac{1}{4}$, $\frac{2}{3}$, 2, –2.5, –4

22. Slopes are positive in Exercises 1, 3, and 4. Slopes are negative in Exercises 2, 5, and 6. Positive slopes slant up to the right. Negative slopes slant up to the left.

23. Answers may vary.

Lesson 18 Graphing Systems of Linear Equations

1. (2, 0) 2. (4, 0)
3. (1.5, 1.5) 4. (1, –1)
5. (1, 3) 6. (0.5, 1.25)
7. (–2.4, –1.8) 8. (3.6, –1.1)
9. (4, –1) 10. (0.5, 2)
11. (1.5, 3) 12. (–1, 2)

13. Slope = $\frac{1}{2}$; $y = \frac{1}{2}x + 2$

14. Yes at (–1, 1.5) because they have opposite slopes.

15. $1.69x + 2.29y = 16.52$; $x + y = 8$; Marcus bought 3 packages of the tea that cost $1.69 and 5 packages of the tea that cost $2.29.

16. $4x + 4y = 16.92$ and $2y + 6x = 16.38$; Each soda costs $1.98. Each sundae costs $2.25.

17. Meet at (3, 2).

18. Boat which starts at (0, –1) has further to travel.

Lesson 19 Graphing Inequalities: A Two-Dimensional View

1. $x > 6$

2. $x > -2.5$

3. $x \le 5$

4. $x \ge 6$

5. $x < 4$

6. $x \ge -3$

7. $x < 6$

8. $x > 7$

9. [0 to 675 mi]

10. [0 to 945 mi]

11. 30 mi/gal 12. 810 mi

13. Several methods are possible. 1. Multiply 30 mi/gal by 27 gal. 2. Add 675 mi and 945 mi and divide by 2. 3. Multiply $\frac{1}{2}$ tank (13.5 gal) by city driving and highway driving and add, (13.5)(25) + (13.5)(35).

14. Answers may vary.

Lesson 20 Finding Averages and Drawing Scatter Plots

1. $\overline{y} = 18.97$ 2. $\overline{y} = 1876.875$
3. $\overline{y} = 44.9$ 4. $\overline{y} = 54.83$
5. $\overline{y} = 15.78$
6. Mon: 146; Tue: 147; Wed: 187; Thu: 142; Fri: 149
7. Week 1: 154; Week 2: 153; Week 3: 155; Week 4: 155
8. 154
9. The averages for each week were quite consistent except week 2 was a little less. The average for Wednesday was higher than for the other days. Perhaps a favorite food is served. The cafeteria needs to plan to feed more people on Wednesdays.

10.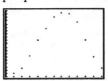

The yearly average is 61° which reflects two months, April and October. Seasonal averages are appropriate: Winter: Dec, Jan, Feb: 44; Spring: Mar, Apr, May: 61; Summer: Jun, Jul, Aug: 77; Fall: Sep, Oct, Nov: 62

11. On an average the passer attempts 25.4 passes and completes 11.9 passes. This gives a 47% (11.9 ÷ 25.4) expectation of competions out of attempts.

12. Answers may vary.

Lesson 21 Exploring Data and Line Relationships

1. $y = 13.66 - 0.169x$; $r = -0.9895$
2. $y = 20 + 0.45x$, $r = 1$
3. $y = 0.69 + 1.32x$, $r = 0.9999$
4. $y = 6.52 - 0.032x$; $r = -0.9572$
5. Exercise 2 has an excellent fit with $r = 1$. Exercise 3 has the next best fit with $r = 0.9999$. Exercise 1 with $r = 0.9895$ and Exercise 4 with $r = 0.9672$ both have a very good fit.
6. $y = 8.49 + 0.65x$; $r = 0.8892$ In theory the measurements should be equal and have a slope of 1. The line of best fit has a slope of 0.65 which is not very good.
7. Answers may vary.